Drones - Various Applications

Edited by Dragan Cvetković

Published in London, United Kingdom

Drones - Various Applications
http://dx.doi.org/10.5772/intechopen.1000551
Edited by Dragan Cvetković

Contributors
Koç Mehmet Tuğrul, Ahmed N. Sayed, Omar M. Ramahi, George Shaker, Ogbonnaya Anicho, Daitao Xing, Anthony Tzes, Rahmi Elagib, Ahmet Karaarslan, Lionel Fouellefack, David L. Kulhavy, Daniel R. Unger, I-Kuai Hung, Victoria M. Williams, Yanli Zhang, Reid Viegut, John-Thones Amenyo, Wolanyo Kpo, Ramasenderan Narendran, Thiruchelvam Vinesh, Soon Hou Cheong, Han Xiang Yee, Hicham Sekkati, Jean-Francois Lapointe, Mikhail Gorobetz, Leonids Ribickis, Anna Beinarovica, Aleksandrs Kornejevs, Yan Guo, Jia He, Jingyi Huang, Xiuzhong Yang, Zhou Shi, Laigang Wang, Guoqing Zheng, Lala El Hoummaidi, Abdelkader Larabi, Khan Alam

First published in London, United Kingdom, 2024 by IntechOpen
IntechOpen is the global imprint of INTECHOPEN LIMITED, registered in England and Wales, registration number: 11086078, 5 Princes Gate Court, London, SW7 2QJ, United Kingdom

British Library Cataloguing-in-Publication Data
A catalogue record for this book is available from the British Library

Additional hard and PDF copies can be obtained from orders@intechopen.com

Drones - Various Applications
Edited by Dragan Cvetković
p. cm.
Print ISBN 978-0-85466-164-0
Online ISBN 978-0-85466-163-3
eBook (PDF) ISBN 978-0-85466-165-7

For EU product safety concerns:
IN TECH d.o.o., Prolaz Marije Krucifikse Kozulić 3, 51000 Rijeka, Croatia,
info@intechopen.com or visit our website at intechopen.com.

Meet the editor

Dragan Cvetković graduated in aeronautics from the Faculty of Mechanical Engineering, University of Belgrade, Serbia, in 1988. He obtained a Ph.D. in Aeronautics from the Faculty of Mechanical Engineering, University of Belgrade, Serbia, in 1997. He has published seventy-one books, scripts, and practicums about computers, computer programs, aviation weapons, and flight mechanics. He has also published numerous scientific papers. Dr. Cvetković is currently the vice-rector for teaching at Singidunum University, Serbia.

Contents

Preface

A drone is an aircraft controlled by a navigator, a pilot with remote transmission of signals from the ground or from another location, or that flies autonomously according to preset memorized data. It should be noted that drones are used for both civilian and military purposes.

The most common use of drones or unmanned aerial vehicles is in the military. Unlike a cruise missile, a drone can be used more than once. It is constructed according to all the rules of the trade like other aircraft, but it does not have a crew or a pilot. A drone designed for one-time use is referred to as a suicide drone, which is constructed cheaply, filled with explosives or other lethal means, and used to destroy critical targets. Disposable drones can also be used as targets for shooting by anti-aircraft defense to train future pilots. Depending on the purpose, drones can be equipped and armed differently, or they can be unarmed. A drone's construction, including its aerodynamics, is identical to that of modern military aircraft and manned helicopters. In fact, drones are more often designed as airplanes and less as unmanned helicopters.

The technique and principles of a drone's takeoff and landing are similar to those of manned aircraft, however, there are also solutions based on appropriate catapulting and landing with a parachute. Usually, the drone is controlled in a combined or hybrid manner. In one part of the path, it is guided from a ground station with remote transmission of the command signal, and in the other part of the path, the drone flies autonomously according to preset navigation data and set components of the flight profile.

Currently, military drones predominantly perform reconnaissance and espionage tasks. Nevertheless, they are becoming increasingly involved in direct combat missions. All armed forces intend to reinforce the number of drones in their arsenal of weapons, with different tasks.

Drones are also used for civilian purposes, such as aerial photography, fire detection and firefighting, monitoring flood-prone areas, traffic monitoring, migration or animal movement monitoring, monitoring of various pipeline routes, and providing appropriate data to video security systems. In addition to surveillance, drones are used in agriculture, to collect data for scientific research purposes (especially in dangerous situations for humans such as collecting data from volcano craters or other dangerous sources), and delivery of products and packages. Recently, drones have begun to be used in certain types of "sports".

Navigator or drone pilot is expected to be one of the most in-demand jobs of the future. Huge amounts of money are being invested in drones and this is an obvious signal that drones are playing an important role in both military and civilian fields.

This book provides a comprehensive overview of drone technology and applications. Chapter 1 provides an overview of drone technologies and related applications.

Chapter 2 discusses the importance of the detection and classification of drones, which is crucial due to their potential use in illicit activities. Radar systems can provide a promising solution when combined with machine learning and artificial intelligence models, as outlined in the chapter. Chapter 3 is dedicated to defining autonomy or autonomous capability from the context of a single drone. However, future drone applications will be used as multi-drone systems consisting of multiple drones. Chapter 4 deals with deformable correlation networks for tracking and segmenting aerial objects, including drones. Chapter 5 discusses the modeling and simulation of a solar-powered quadcopter using the MATLAB package. Chapter 6 is devoted to issues "related" to electric unmanned aerial vehicles or drones. Chapters 7 and 8 are devoted to the integration of drones into the curricula of the respective educational institutions in different environments. Chapters 9–13 examine the application of various drones in specific situations, such as fire detection and warnings, bridge inspection, collision prevention, detection of appropriate elements on agricultural goods, and mapping of agricultural goods based on predefined parameters.

I would like to express my sincere gratitude to all the authors and coauthors for their contributions. The successful completion of this book, Drones – Various Applications, is the result of the cooperation among many people. I especially want to thank Publishing Process Manager Ms. Mirna Papuga at IntechOpen for her support during the publishing process.

Dragan Cvetković
Singidunum University,
Belgrade, Republic of Serbia

Chapter 1

Drone Technologies and Applications

Koç Mehmet Tuğrul

Abstract

Unmanned aerial vehicles, previously used for military purposes, have started to be used for civilian purposes since the 2000s. With the widespread use of unmanned aerial vehicles, mostly used for defense purposes, they have turned into flying vehicles called drones. Today, it is used for different purposes such as taking pictures, taking images from inaccessible places, competitions, and having a good time. However, it continues to evolve, playing an important role in social media and inter-company competition. In recent years, studies have continued on small drones that can fly autonomously, especially in closed areas, and assist people. The shrinkage and cost reduction of electronic components such as microprocessors, sensors, batteries, and wireless communication units shows that drones smaller than 1 kg will soon occur in many areas of our daily life at much more affordable prices. This rapid change, development, and proliferation of drones also raise privacy and security concerns. While obtaining images of people and institutions in supposedly private environments with such technological devices emphasizes individual security, its use in urban areas and airports draws attention to social security. The rapid developments in this area necessitate the necessary regulations and controls to prevent accidents and other problems due to the growth in use and loss of control.

Keywords: unmanned aerial vehicle, drones, drone features, drone technologies, drone usage fields

1. Introduction

Although the original translation refers to a flying electromechanical technology used to mean "drone," the term drone is based on a game using the term "Queen Bee" [1]. The historical development of drones shows that the first vehicle that fits the definition of unmanned aerial vehicles (UAV) was the unmanned hot air balloon used in France in 1783. Since then, drones have continued to be used in intelligence, aerial surveillance, search and rescue, reconnaissance, and offensive missions as part of the military Internet of Things (IoT). Drones are widely used in many areas, such as traffic surveillance, cargo, first aid, agriculture, entertainment, hobby, security, and surveillance, as they provide many customizable solutions that combine practicality and speed.

In any case, the low resolution of open-source images provided by satellites, the expensiveness of high-resolution images, and the dependence of satellite images on weather conditions create significant problems in the supply and evaluation of images. Drones fill an important gap as images taken by drones that can fly several meters above the ground offer advantages such as cloud-based data analysis, allowing manufacturers to monitor product development and quality continuously, easily, and quickly [2].

Such technological equipment also allows real-time monitoring of the business situation in the construction industry, rapid analysis of the excavation area in the mining industry, precise determination of the excavation to be excavated, and preliminary preparations. In the energy and infrastructure sector, it is possible to determine roads, cables, and pipelines and plan accordingly. Aid organizations use drones to locate camps, plan transportation routes, and monitor work. It allows the rapid delivery of goods and services and the arrangement of communication infrastructures to areas with a high density of buildings and people or where there is no highway transportation.

Drones for transportation fulfill important tasks in delivering medical supplies and foodstuffs over long distances in emergencies and rapid rescue efforts. Drones are also used in the logistics industry to detect damage and cracks in the ship structure and hull, allowing emergency teams such as the fire brigade to intervene in dangerous areas quickly and safely. We also see drones in measuring the level of wear and tear on highway routes, security checks in bridges and tunnels and other determinations, and interior controls of partially damaged buildings in disaster. To provide the communication needs of different work groups or rescue teams in the field of activity, drones can be used in communication as well as helping to establish private communication networks quickly. Although the battery life problem, the biggest obstacle in using drones, is still a problem in long-distance tasks, it is possible to perform longer functions by changing the battery in short-distance studies. Another important limitation of unmanned aerial vehicles used today is the need for human supervision to perform almost all of the tasks described, which are factors that slow down the performance of the specified functions and the intervention in case of danger.

Many factors are considered depending on the size and flight characteristics of drones. Among these, the increase in energy cost due to the reduction in size and the difficulty of hanging in the air can be counted. On the other hand, there is no ideal design for fixed and rotary wing drones that combines both aerodynamic and propulsion performance. These include traditional fixed-wing and rotary-wing designs and bio-inspired designs based on flapping wings. Of course, each of these designs has advantages and disadvantages. For example, fixed-wing aircraft can fly quickly and efficiently but cannot hover. Rotary-wing designs can hover and are highly maneuverable but have lower flight efficiency. On the other hand, there is no ideal design for both fixed and rotary-wing types that provides both aerodynamic and thrust performance [2].

The growth in the use of commercial and personal drones has necessitated many regulations to prevent accidents and provide drone control in a way that does not pose a hazard [3]. Although many countries have created UAV regulations, the increasing use of drones causes rules to change constantly and new rules to be introduced. These regulations vary between countries and regions. Among these regulations, in the UK, the Civil Aviation Authority (CAA) limits the flight altitude of unmanned aerial vehicles to 500 feet, drones weighing more than half a kilo. It is important to register it with the CAA. The agency also states that it is prohibited to fly near airports and an aircraft with its "Dronecode" that it is mandatory to stay below 400 feet and at least 150 feet away from buildings and people, and that the drone must be constantly observed during flight [1].

DOI: http://dx.doi.org/10.5772/intechopen.1001987

2. Features of drones

Drones, designed with a kite-like mechanical architecture and one engine placed on its diagonal points, mainly consist of four basic components: propeller, engine, body, and flight board. Drones, generally defined as 250 and 450 class according to the distance between the two crossed motors, are classified as single rotor (helicopter), multi-rotor (multicopter), fixed wing, and fixed-wing hybrid VTOL according to their physical structures.

Single-rotor drones are small-sized helicopters and fueled or electric types are available. Working with a single blade and fuel offers advantages such as increasing stability and flying longer distances, but it also brings safety risks.

Multicopter are the smallest, lightest, and most widely used drones on the market. Their flight distance, speed, altitude, and payload are limited. These drones, which usually carry a light load such as a camera, are used for terrestrial observations and determinations of up to 50 minutes of flight time.

Multicopters are divided into models with four engines (quadcopter), six engines (hexacopter), and eight engines (octocopter) according to the number of motors. The main factor determining the design here is the carrying capacity of the drone and the required range; accordingly, the size and number of engines are determined. According to the necessary range and control structure, there are many types of drones, from hobby-purpose models with a range of 30–40 meters to professional models with a range of 10 km+, together with the engine, chassis, and battery.

- Tricopter: It is a type of drone that can take off and land vertically and has six degrees of freedom on the X–Y–Z axes. Its cost is lower than other options. However, it is an important disadvantage that it is not symmetrical (**Figure 1**).

- Quadcopter: It is the most preferred type of drone. It is simple as well as versatile. In this way, it provides ease of use. It has four propellers and four motors, two on the right and left, and can take off with a load of up to 5 kg. A higher flight comfort can be achieved with its four-arm structure being symmetrical. However, since the quadcopter system works as a whole, it has no backup plan or redundancy. Therefore, in case of any malfunction, the drone will most likely crash (**Figure 2**).

- Hexacopter: It is a type of drone with six propellers. It is a type of drone that can offer excellent performance even on indoor flights, thanks to its advanced forward vision system and ultrasonic sensors. It can be equipped with various

Figure 1.
Tricopter.

Figure 2.
Quadcopter.

equipment and take off with a load of up to 10 kg, as in other models. It has the features of using the vehicles in automatic or manual mode with products called desk or vehicle terminal, which enable longer range operation outside the flight control system and hand terminal, such as returning to the starting position when it goes out of the coverage area, when the signal is cut off or when the battery is about to run out (**Figure 3**).

- Octocopter: It is a type of drone with eight propellers. It is an advanced type of drone that can take off with a load of 25 kg with its equipment. It is especially preferred for heavily loaded works (**Figure 4**).

- Fixed-wing drones: Unlike rotary wings, they use wings like a regular airplane instead of vertical lift rotors to provide lift (**Figure 5**). In other words, they only need to use energy to move forward. They are much more efficient as they do not use additional power to stay in the air, so they can cover longer distances and scan much larger areas. Gasoline-powered ones can remain in the air for 16 hours or more.

The main disadvantage of a fixed-wing aircraft is that they cannot fly in one spot. Depending on their size, they need a runway or launcher to get them into the air. Fixed-wing also makes launching and landing them much more difficult, as a runway, parachute, or net may be needed for a safe landing again after the flight. Only the smallest fixed-wing drones are suitable for manual launch and "belly landing" in an open area. Other negative aspects are the high cost and difficulty of using fixed-wing

Figure 3.
Hexacopter.

Figure 4.
Octocopter.

Figure 5.
Fixed-wing drone.

drones. Training and certification are needed to develop control capabilities to enable takeoff, flight, and soft landing when a fixed-wing drone is launched. The difference to be considered by the drone pilot is that a fixed-wing drone always moves much faster than a forward, multi-rotor drone.

- Fixed-wing hybrid VTOL drone: Fixed-wing hybrid VTOL drones are a new hybrid category that can also take off and land vertically, combining the advantages of fixed-wing UAVs and the ability to hover. These drones are a mix of fixed-wing drones with rotors attached to the blades and rotor-based drones (**Figure 6**). There are fixed-wing designs with no movement feature of vertical lift engines, models with motors that provide forward motion, and various types with propellers that can move up and down for takeoff and horizontally for forward flight. Fixed-wing hybrid VTOL unmanned aerial vehicles, thanks to its hybrid approach, this technology offers users the durability of a fixed-wing design and the vertical flight capabilities of a rotor-focused design. With the introduction of modern autopilots, gyroscopes, and accelerometers in this field, it has become easier to operate these drones.

Fixed-wing hybrid VTOL unmanned aerial vehicles have the advantages of flying vertically and hovering, offering greater versatility than fixed-wing drones and the

Figure 6.
Fixed-wing VTOL drone.

durability needed to travel long distances with heavy payloads. On the other hand, they could be better in both forward flying and hovering features, and the need for specialized personnel in the flight and control of these types of drones is an important consideration.

Drones, divided into four main groups according to wing type, are used in aerial mapping, inspection, surveillance, agriculture, and search-and-rescue areas. Drones can be classified according to size, from very small to large drones [3].

Drones size, very small drones, length of 150 mm, weight of 200 g; small drones, length of 151–300 mm, weight of 200–1000 g; medium-sized drones, length of 300–1200 mm, weight of 1–20 kg; large drones length of 120 cm, weight is classified as 20 kg. The payload a drone can carry will depend on the engine's power and the lift generated by the propeller in standard weather conditions.

According to the payload they can carry, drones are featherweight, with a weight of less than 11 g and a load of 4–100 g. drones, light drones with a weight of 200–1000 g, a payload of 150–270 g, medium-weight drones with a weight of 1–600 kg, a load of 400–1460 g Drones, and heavy-lift aircraft with a weight of more than 160 kg and a payload of more than 1000 kg are divided into four classes. Featherweight drones are mostly used for military surveillance, light and middleweight drones are used in recreation and photography, and heavy-lift aircraft are used in the cargo transport and cinema sector.

Drones by range, very close range drones, 5 km range and 1-h flight time; close range drones, 50 km range, 6-h flight time; short-range drones, 150 km range, 12-h flight time; medium range drones, 644 km range, 24-h flight time and long-range drones are divided into more than 644 km range and more than 24-h flight time. The drones in this grouping are mostly used in military surveillance, aerial photography, and mapping works.

For a drone to fly, it must have a power source such as batteries or fuel. Their power sources classify drones as battery-powered, gasoline-powered, hydrogen fuel cell, and solar drones. Despite the advantages of battery-operated drones, such as lightweight, sufficient energy storage, and high discharge rates, they have disadvantages, such as short lifespan, rapid energy consumption, and burn risks. Gasoline drones have important advantages such as not needing expensive spare batteries and charging stations, no need to wait for the storms to recharge, high flight speeds, carrying heavy loads, and long flight times. On the other hand, it has undesirable features such as large size, noisy operation, and combustion and explosion due to fuel. Although hydrogen fuel cell drones have positive features such as using renewable and environmentally friendly energy, higher energy efficiency than batteries, long flight time, and short refueling, they have disadvantages such as excessive heat generation and low operating efficiency. Solar drones, on the other hand, have low operating costs and are lightweight but have limited flight time.

According to the motor type, drone types are divided into brushed and brushless. Brushed motors are the type of motor used in recreational drones. Although these engines are inexpensive, they require much maintenance. The larger the dimensions of these engines, the greater their power. Most motors are not in direct contact with the propeller. Brushless motors, being in connection with the rotor, reduce the need for maintenance and increase operating efficiency. It has much higher power than brushless motors. However, since this type of drone motor predominantly works with alternating current, each motor must have an electronic speed controller (ESC) that adjusts its rotation speed [3].

3. Main structural elements of drones

3.1 Frame

The drone body is the main skeleton system where all drone parts are placed (**Figure 7**). The drone body can be compared to the chassis in computers. The propellers, motors, battery, camera, and receivers fit into the fuselage and are ready to do their job. It has an "X" style design with four arms extending from the middle body. All other parts are in the body, especially the rotors, battery, cards, and camera. If we compare it with the human anatomy, the body forms the skeleton of the drone. F330, F450, and F550 models are the most preferred drone bodies. The most important criteria in choosing these bodies are the ease of spare parts supply, suitability of the price/quality ratio, and the performance it provides. Despite its high price, the QAV250 fiber carbon material is another preferred drone body because the carbon-fiber material reduces the possibility of breakage in case of falling [4].

Figure 7.
Drone frame.

3.2 Propellers

Also called rotor or blade, drone propeller comes in various shapes, sizes, and materials. Regardless of their specific characteristics, all propellers are the basic equipment that does not allow the drone to take off (**Figure 8**). The propeller is one of the most sensitive parts of the drone and can be easily damaged as it needs to be constantly replaced or installed. These parts, which can be easily dislodged, broken, or corroded during the flight, must be checked. The propellers should be maintained for a good and uninterrupted flight, and continuous preflight checks should be carried out. Propellers are generally made of plastic material, and shaft and propeller diameter are important parameters when changing.

3.3 Motors

The drone motor is the main dynamo that makes the propellers spin and provides enough thrust for flight. They convert the electrical energy stored in the battery into motion energy and enable the propellers to rotate (**Figure 9**). Motors, one of the most important parts of the drone, are very important, especially for the drone to work effectively. Removing dust and debris is necessary to prolong motor life and ensure trouble-free operation. If the engine makes an unusual noise, it is time to check the motor. Two types of brushed and brushless motors are used, which vary according to their size, weight, and load. Brushed motors are preferred for lighter and partially entry-level drones, and brushless motors are preferred for heavy and professional drones. While determining the drone motor, size, work, and speed are important parameters.

3.4 Battery

The battery, described as the heart of the drone, is the most important part in terms of power generation and performance (**Figure 10**). The lithium polymer or lithium-ion battery inside can easily supply propellers' movement energy. Drone batteries are defined as 1s, 2s, and 3s according to the number of cells in the battery. Each cell corresponds to a voltage of 3.7 volts. In this case, 1-cell lithium polymer batteries are preferred in light hobbyist drones, while 2- or 3-cell batteries are used in drones that require higher voltage. They are not easily damaged, but they can be discharged. For this reason, uninterrupted flights need to have spares for batteries and propellers.

Figure 8.
Propellers.

Figure 9.
Motor.

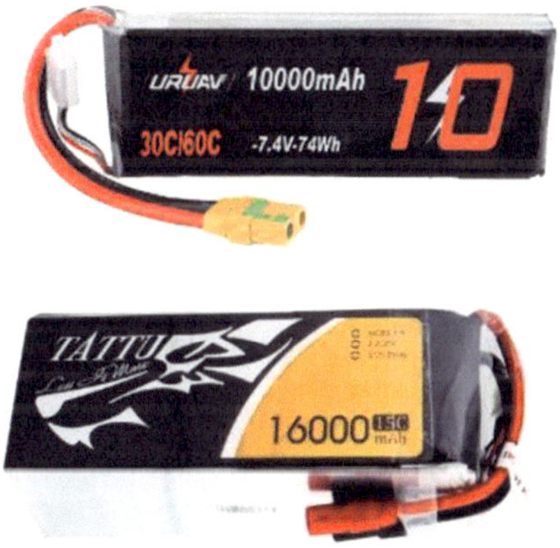

Figure 10.
Power supply.

3.5 Flight control cards

This part, which provides the necessary controls by interpreting the signals from various cards, sensors, and receivers, is the drone's brain. The flight control card is a circuit board that enables the UAV to move through the sensor information it receives. Sensors are just like a nervous system and are essential for boarding cards. It converts received signals into action for setting speed and direction, activating the camera, and executing other commands. By combining many sensors such as the gyroscope, magnetometer, and accelerometer with GPS, flight control CPUs can use data about the UAV's status, speed, position, and direction to guide the vehicle's flight and operation according to parameters. These operations are guided by the settings determined by the user before the flight. Other sensors also enable the autopilot to avoid obstacles and collisions while operating. Autonomous UAVs, on the other hand, can perform all flight plans, including VTOL or runway takeoffs, in-flight maneuvers, and landing, according to defined parameters (**Figure 11**).

Figure 11.
Flight control board.

3.6 Power distribution board

PDB (Power Distribution Board), mounted on the Drone; the circuit board organizes the power connections of batteries, ESCs, and other onboard systems. It is unnecessary for all drones but helps create a neat and organized drone (**Figure 12**). In particular, it is responsible for power distribution from the flight battery to each electronic speed controller for the drone control board. In some cases, the PDB is also used to power components such as the camera, LED taillights, and flight controller.

3.7 Electronic speed controller

Like the PDB, the Electronic Speed Controller (ESC) is another essential part that only requires a little attention or maintenance. Its task is to control the speed of electronic motors (**Figure 13**). Each motor has its own ESC connection. In many standard installations, each branch has a separate ESC. However, 4-in-1 ESC parts are also available and provide different advantages. The ESC is the first place to look when a power-related problem occurs, especially local issues with a particular engine.

DOI: http://dx.doi.org/10.5772/intechopen.1001987

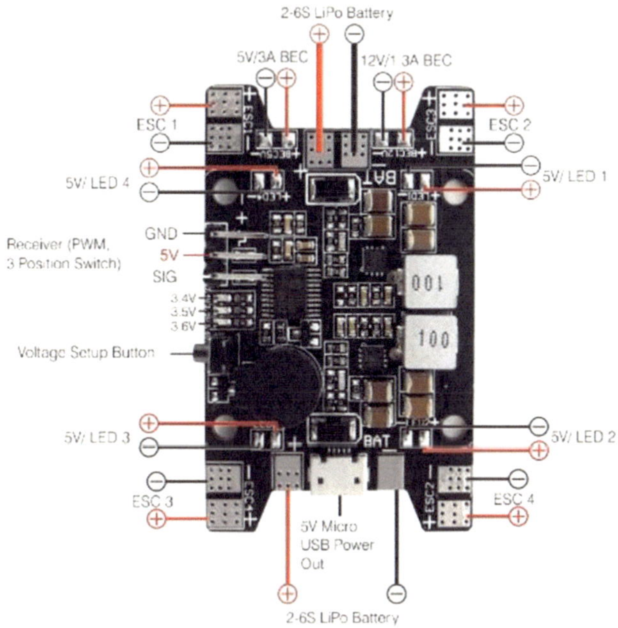

Figure 12.
Multiple power control board used for quadcopter.

Figure 13.
Electronic speed control and controller.

3.8 Drone controller and receiver

The drone controller is the tool for managing and directing the drone. The most important feature of the remotes is the number of channels. The number of channels should be as much as the type of command given to the drone. While 6-channel controls are mainly used in drones today, this number can increase with accessories such as cameras. A digital display on the controls is another feature that makes it easier to use (**Figure 14**).

Receivers are like sensory nerves; they collect information (signals) from external sources (radio transmitters) and transmit them to the brain (flight controller). It is

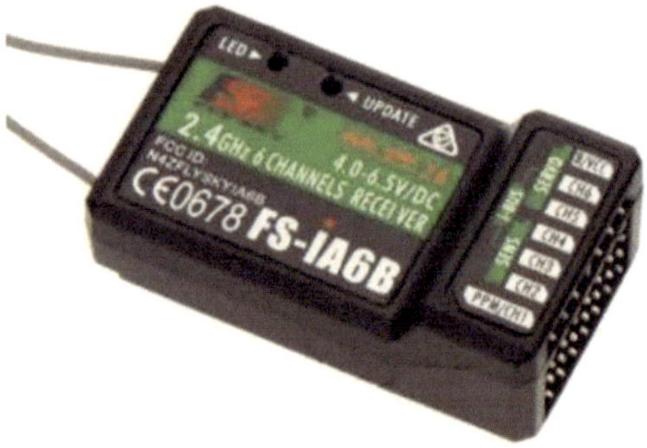

Figure 14.
Six-channel drone receiver.

the part that detects the commands from the remote control and enables the drone to move. Receivers using radio waves offer a very fast and uninterrupted transmission. Typically, they use an integrated antenna that allows the reception of radio signals. The receiver will have four channels to control the down, oscillation, acceleration, and yaw. However, many receivers have auxiliary channels to perform certain functions or trigger flight modes. Five or 6-channel receivers are generally the preferred models [4].

3.9 Camera

Not all drones have a camera system, but models offer additional options in-flight, such as aerial photography, cinematic photography, or pilot's eye view (FPV) flight. As with drones, video systems are available in various models and layouts when considering drone uses and have their parts, such as the gimbal, video transmitter, antenna, and FPV goggles (**Figure 15**).

One of the most important components in FPV (first-person view) imaging is the FPV camera. The FPV camera is mounted in front of the aircraft, and images from the camera are transmitted to the pilot's glasses or monitor. There are two types of FPV cameras, analog and HD [5]. In these drones, the image comes first to the lens on the FPV camera; the sensors capture the image and convert it into electrical signals. These converted signals are sent to the video transmitter. When choosing a camera,

Figure 15.
Camera.

parameters such as sensor type (CCP/CMOS), sensor size, dynamic range, width ratio, field of view (FOV), delay, and low-light performance are the features to be considered.

There are two sensors used in cameras as CCD and CMOS. CCD sensor collects data using scanning method pixels simultaneously. It has less jelly effect due to the global more shot, high dynamic range performance, good performance in multi-light and low-light (WDR) environments, and better black/white transitions than CMOS. CMOS sensors collect line-by-line data for all pixels on the horizontal and vertical axis. This process can create delays, according to CDD. CMOS sensors may generally experience image distortions with the rolling shutter, that is, the jelly effect, which occurs when the CMOS sensor collects data line-by-line from the pixels. These sensors have good color fidelity and performance, have low power consumption, and are cheaper.

Another factor affecting image quality in cameras is sensor size. Low-light performance and dynamic range depend on sensor size. The larger the sensor size, the greater the field of view as the lens size and low-light performance increase. The dynamic range improves the image in bright and dark environments according to the light intensity and allows the user to see the desired object comfortably. The cameras have two aspect ratios, 16:9 and 4:3. Many cameras today have an adjustable double-width balance. Field of view (FOV) and angle are important user features. The larger the angle, the more difficult it is to view distant objects due to the fisheye appearance. The smaller the field of view, the closer and clearer the image. This event is preferred on immediate flights.

On the other hand, decreasing the lens length or increasing the sensor size increases the field of view. The "lux" unit is an important feature when shooting in low-light environments, and the increase in this value is a feature that improves low-light performance. Sensor size is very important in low light. As the sensor grows, the sensor area increases, causing more light to enter, and lower speed performance occurs.

3.9.1 Sensor features of drone cameras

Drones, cameras, and sensors detect differences, modeling, classification, development, and changes, especially for agricultural applications. Imaging sensors such as red, green, and blue (RGB), and near-infrared, multispectral, hyperspectral, and thermal and distance sensors such as LiDAR and search-and-rescue (SAR) are widely used. RGB sensors capture the image as the human eye sees these red, green, and blue (RGB) colors, a narrow band of the electromagnetic spectrum.

Multispectral image sensor offers the opportunity to easily detect differences in the target area by using sensors sensitive to certain wavelengths along the electromagnetic spectrum. The sensor types and features used for this purpose are divided into green, red, and blue visible bands, red edge, and near-infrared.

Visible light has wavelengths in the range of 400 to 700 nm. It determines regional features, elevation modeling, and object counting applications, especially for agricultural uses.

Red Edge: The 717 nm center is a band corresponding to the 12 nm bandwidth. This tape provides information on phytosterols and chlorophyll. Accordingly, it is used in plant health, plant counting, and water management.

Near-infrared (NIR): 842 nm center is a band corresponding to 57 nm bandwidth. This reflection is used in soil, moisture analysis, crop health, and stress analysis depending on the chlorophyll level in the plant.

Like other spectral imaging, hyperspectral imaging collects information and processes it into the electromagnetic spectrum. Nevertheless, aside from visible light, which the human eye can detect in three bands (red, green, and blue), spectral imaging looks at objects using a wide part of the electromagnetic spectrum. In other words, thanks to this technique, which divides the image into many bands, it offers the opportunity to grasp objects and their properties in a much wider band range than what is visible in the pictures with a single camera. In particular, it is a technology that can be used in the detection of underground resources in mining, agriculture, the prevention of diseases and pests, the military field, thermal infrared hyperspectral imaging, the chemical field, in the detection of colorless and odorless harmful substances in the air, in environmental issues, in the detection of leaking toxic wastes. Despite the advantages of hyperspectral imaging, such as imaging in a wide spectrum, it is very expensive, and complex processing processes pose a significant problem [6].

Thermal imaging is an imaging system based on invisible IR energy (heat) and determines the general structure of the image, colors, and shapes formed according to IR energy. While normal cameras create the image thanks to the light, thermal cameras make the image thanks to the heat. Similarly, color differences are important when the human brain and eye use colors and light to create an image.

Thermal cameras are used to map the amount of water in the soil (SWC) depending on the land surface temperature (LST) [7]. Thermal cameras have limited spatial resolution, which often causes difficulties in homogeneous areas such as farmland with bare soil [8]. It is especially used in agriculture to determine plant water needs, detect disease, and for phenotyping.

4. Drone technologies

4.1 Radar positioning and returning home

Today's drones can operate in GNSS and nonsatellite modes using a combination of Global Navigational Satellite Systems (GNSS) such as GPS and GLONASS. This precision navigation system brings 3D map creation, navigation, and search-and-rescue (SAR) tasks while flying. Satellites take advantage of GNSS satellites when the drone is first powered up. It uses constellation technology. A satellite constellation is a group of synchronized satellites with overlapping coverage areas. The built-in compass lets the drone and remote control system know exactly the flight location. The compass is calibrated to set a starting point. Thus, the drone's position will return in case of signal loss between the drone, and the remote control system is determined. This adjustment is also known as the "fail-safe function." The radar technology in the drone checks some information on the control screen before starting the flight, confirming the suitability of the current conditions for flight. This information includes that communication with good drone GNSS satellites has been established, and the drone is ready-to-fly. It provides information such as determining the current location, returning home using the button for any reason, reaching the set low battery level, and recording the starting point for the "Return to Home" feature in case of loss of communication between the drone and the remote control, and can return to the starting position without any problems [10].

4.2 Obstacle detection and collision avoidance technology

Drones are equipped with collision avoidance systems for safety purposes [9]. These systems consist of sensors that do not scan the environment but detect obstacles. Data from the vision sensor, ultrasonic, infrared, LiDAR, time-of-flight (ToF), and monocular vision sensors transform images into 3D maps, allowing the flight controller to detect and avoid the object. With the help of software algorithms and SLAM technology using this data, 6-way obstacles are detected, and necessary actions are taken to prevent them.

4.3 Drone movement patterns

Roll is the Z-axis on the nose and tail of the aircraft, i.e., longitudinal. It means it rotates around the axis; the wings move up and down (**Figure 16**).

Pitch is the X-axis between one wing and the other wing of the aircraft, Lateral. It rotates around the axis; in other words, the nose and tail of the aircraft move up and down is also called the pitching motion.

Yaw is the Y-axis that runs between the upper and lower parts of the fuselage, passing through the aircraft's center of gravity. It means the rotational movement on the axis. In other words, it is the right-left movement of the nose and wing of the aircraft [8].

4.4 Gyro stabilization, IMU, and flight controllers

The gyroscope immediately responds to varying forces coming into the drone by providing the necessary navigation information to the central flight controller. The inertial measurement unit (IMU) detects the current acceleration rate using the accelerometer and the gyroscope to see changes in rolling characteristics such as pitch, roll, and yaw. The IMU includes a magnetometer to calibrate against drift during flight. The gyroscope and IMU are core components of the drone flight controller, which is the drone's brain. Gyro stabilization technology is one component that provides the drone with smooth flight capability by using all these components [10].

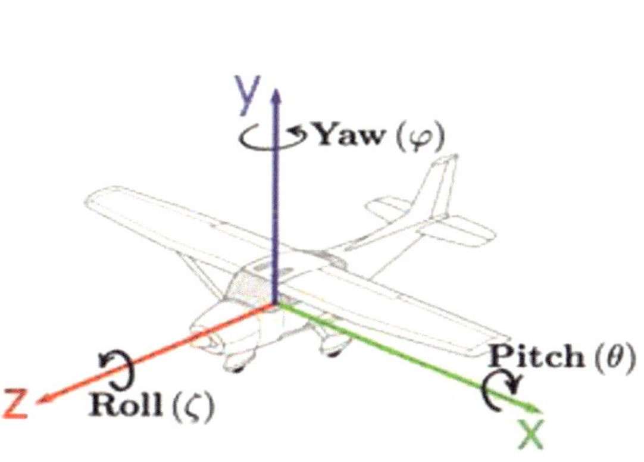

Figure 16.
Drone movements.

4.5 Drone engine orientation and propeller design

Motors and propellers are the technology that lifts the drone, allowing it to fly or hover. On a quadcopter, the motors and propellers work in pairs, with two motors/propellers rotating clockwise and two motors/propellers rotating counterclockwise. The drone moves according to the data it receives from the flight controller and electronic speed controllers (ESC) for hover or engine flight direction.

4.6 GPS ready-to-Fly and No-Fly zone drone technology

According to the compass calibration, when the drone starts to work, the drone searches for the position of GPS satellites; when more than six satellites are detected, the drone becomes "ready-to-fly." To increase flight safety and prevent accidents in restricted areas, the "no-fly zone" feature is activated, and flight is blocked. No-fly zones are divided into two categories, A and B. These regions can be renewed and changed by software updates. Suppose all controls are in the direction of flight. In that case, the flight of the drone is provided, and the existing flight telemetry and the objects or environments that the drone monitors with the camera can be followed over the mobile device or the control device.

4.7 Creating 3D maps and models

The drone visualization's hyperspectral, multispectral, LiDAR, and thermal sensors are created by 3D digital surface models (DSM) of buildings, earth, and landscape and the land's digital height maps (DEM). It has become very easy to make evaluations and plans about agricultural products, flowers, fauna, shrubs, and trees.

Time-of-Flight (ToF) is a precision mapping and 3D imaging technology that emits a very short pulse of infrared light and measures the rotation time of each camera sensor pixel. Flash LiDAR time-of-flight cameras offer a variety of solutions by measuring distances within a captured scene. This way, operations such as object scanning, obstacle avoidance, tracking and recognizing objects, and measuring volumes become easier.

Turning images into maps and models is as important as taking high-resolution drone images. For this, Pix4D Mapper, AutoDesk ReCap, 3DF Zephyr, Agisoft Photogrammetry software such as PhotoScan and ESRI Drone2Map for 2D/3D mapping and imaging software such as ArcGIS, PrecisionHawk, and DroneDeploy are among the most widely used tools.

5. Systems supporting the development of drone technology

The use of new technologies with existing systems is an important issue that ensures the development of the system and the increase in performance. With the addition of new generation technologies such as artificial intelligence (AI), the Internet of Things (IoT), big data, blockchain, 3D printing, robotics, gene studies, 5G, nanotechnology, and solar cells, which gained the first remote control feature with the use of radio signals, the usage area of drones expanded [11]. When evaluated in terms of the contributions of the developed technologies to the drones, the body, propeller, and engine parts are modeled with a 3D printer; with artificial intelligence technology, the range is increased with a solar cell, and it can process big data quickly with 5G, and control and coordinate the use of a robot for special purposes with the results it produces. A drone that can fly is notable for most industries. The capabilities and usage areas of drones will develop depending on the technology development. Although drones are seen as the end users of developed technologies, they have the potential to be an important tool in the development of technological innovation, depending on the needs.

6. Drone usage areas

Although the use and development of drones are always for military purposes, the use of drones in both commercial and other sectors is becoming widespread today. The evaluations regarding the usage areas still used in this regard and expected to become more widespread soon are summarized below.

6.1 Agriculture

In agricultural areas, it contributes significantly to providing data to sector stakeholders and increasing producers' productivity by collecting data with regular land observation. Drones in spraying, fertilization, and plant damage detection applications inspect the planted crops (**Figure 17**). The use of drones in agricultural production is becoming more common daily, and it provides convenience for producers to optimize production by increasing efficiency and reducing physical load. Drones offer significant time savings in agricultural research, planting seeds, monitoring livestock, and predicting crop yields. Smart farming techniques will become more widespread with drone and satellite data for producers to monitor their products and plan for planting, fertilization, and spraying times.

6.2 Environment

Depending on the increasing city population, drones are successfully used in environmental control and emergency response processes. To prevent environmental pollution, drones not only carry out projects aimed at cleaning the seas but also make an important contribution to the fight against poaching and the tracking of endangered animals. The behavior and disease conditions of animals can be followed thanks to thermal cameras. In addition, oil companies use drones for inspections of oil and gas leaks. Drones with thermal cameras perform important tasks in detecting leaks quickly and preventing possible risks.

Figure 17.
Agricultural drone.

6.3 Health

The use of drones for medical purposes is used to transport response equipment to the scene in urban areas, to direct people to the stage with drones with cameras and audio devices, to transport equipment such as medicine and trauma kits in rural areas, or to facilitate search-and-rescue efforts.

In patients with shockable rhythm in prehospital events, delivering a life-saving defibrillator (AED) by drone is 32% faster in urban centers and 93% faster in rural areas where other vehicles cannot reach [12]. In the same way, it has been demonstrated that it is possible to transport drugs/tissues, and blood products, with drones, with the research conducted with samples taken before and after the flight on the effect of vibration on the transported tissue. These materials can be delivered to difficult areas faster than the traditional methods [13].

6.4 Photography and cinema

Professional video shoots are made today using drones in commercials, TV series, and movie sets, successfully capturing specific images. It is used in the direct marketing of products and in taking aerial images to show a city, beach, or building from a bird's eye view in advertising shoots. Advertising drones are becoming increasingly common in crowded events and areas with high density. It provides great convenience in collecting images and information from places that cannot be visited or entered, especially due to security problems. The use of drones can make a significant contribution to innovation and quality in the film industry.

6.5 Mapping

Drones, becoming increasingly widespread in mapping, can map almost all terrains quickly and in three dimensions. For this purpose, LiDAR Drones with sensors provide highly successful and accurate data. LiDAR technology offers important solutions in the evaluation of agricultural products as well as the mapping of landforms (**Figure 18**).

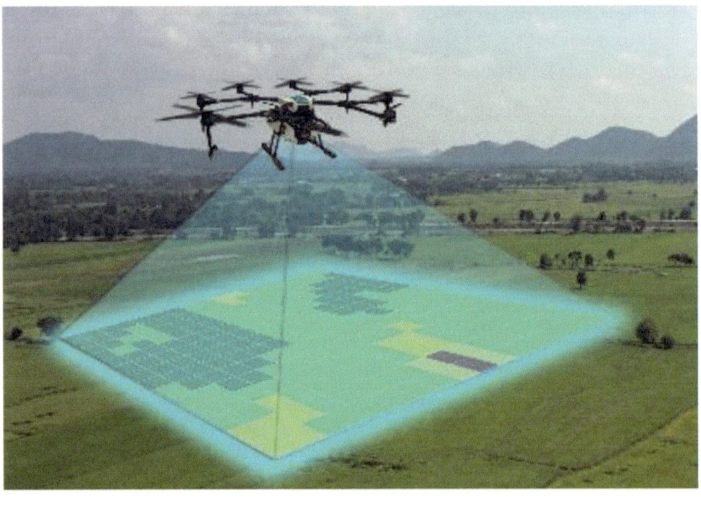

Figure 18.
Mapping with a drone.

6.6 Logistics

Drones are used in the logistics industry to transport food, packages, or goods. It is preferred for transporting urgent or frequently sent small parcels and for delivery to hard-to-reach areas. One of the most important and pending problems in logistics use is that drones need a sufficient carrying capacity. Similar to agriculture, it is also used to scan different warehouse materials. Heavy-duty drones can improve road traffic by replacing existing carriers for material management and transporting goods between warehouses. Soon, drones may be an important player in the delivery options of packages (**Figure 19**).

6.7 Emergencies

Another area where drones are used is in emergencies. Drones help the fire brigade, police officers, volunteer rescuers, and more. Drones are used for reconnaissance before guiding the rescue team, especially when the extent of the disaster is unknown. Autonomous underwater vehicles (AUVs) prevent possible drownings and assist in rescue in situations such as when a boat capsizes. Drones are also used to search for the injured in avalanche cases.

6.8 Conservation of wildlife and historic buildings

Drones are an economical and effective alternative to monitor wildlife species and protect natural life. Rapid aerial observations are important tools in gaining a better idea of the health of the species and ecosystem, monitoring and researching migration routes by monitoring animal groups, and preventing poaching.

Drones also scan forest floors destroyed by fires, releasing seed containers containing seeds, fertilizers, and nutrients to help trees regenerate. Considering that

Figure 19.
Logistics drone.

reforestation of such areas with human labor will take many years, the importance of using drone technology becomes apparent.

On the other hand, drones also play an important role in historic preservation efforts. Drones provide great convenience in creating 3D maps of historical places or structures. The 3D images used to reconstruct lost areas provide clues to experts about culture and architecture [14].

6.9 Military

The first known use of drones was for military purposes. It is used for different purposes ranging from military, espionage, radar system, area detection, and observation to transporting food, weapons, and ammunition. Highly sophisticated drones are used, equipped with tools for thermal imaging, laser rangefinder, and even air strikes. MQ-9 Reaper, one of the drones used for military purposes today, has a length of 36 feet and is equipped with a 1852 km flight system at an altitude of 50,000 feet [14].

7. Barriers to drone technology

Although drones have multiple potential uses, there are certain obstacles to their widespread use. Some of these barriers are summarized below.

7.1 Power source

One of the biggest problems for these vehicles, which have managed to work with increasingly smaller and lighter components, is the power requirement. Due to the limited flight times, developing drones capable of long flights with smaller and more powerful batteries remain the priority to solve this problem. Another important expectation is the development of systems that can analyze the terrain with object detection and recognition software instead of onboard systems that require increased weight and more power.

7.2 Security

The biggest obstacle to drone use is security. All human-operated aircrafts inherently pose a risk of falling. Working on limited battery power, having fast spinning propellers, and the potential to fall from heights greatly threaten living things, structures, and the environment due to the increasing number of drones. In addition to obtaining the necessary permits for flights to be made in residential areas, it is essential to develop technologies to prevent possible accidents by autonomously detecting drone obstacles. The importance of safe use is increasing, especially with using drones in daily life. On the other hand, law enforcement should be put in place to limit drones' interference with the privacy of others.

Drones are becoming a bigger target for cyberattacks as their use increases. Hackers can intercept the transmitted data to take control of the drone. For this purpose, extra measures should be developed to protect drones and the information they store, such as using protective software or passwords.

8. Future prospects

Drones, which have high technology and raise the level of technology daily, are vehicles made of lightweight composite materials that increase strength and maneuverability. Today, drones are used in a wide variety of fields, such as construction, defense, photography, marketing, delivery, agriculture, rescue, and entertainment, and they will certainly meet with drones in new sectors soon, depending on the need for use.

It seems to be a handicap that fixed-wing drones, capable of long-distance flight, cannot perform vertical takeoff and landing. For this reason, long-distance drones with VTOL technology that can take off vertically, hover, and land will become widespread. Drones with improved navigation technology and enhanced connection sensitivity to the group navigation satellite will further improve their ability to direct to the right target. With advanced obstacle detection technology and sensing sensors such as ultrasonic, vision sensor, infrared, LiDAR, and monocular vision, drones will gain new capabilities in 3D mapping. It seems possible to further improve the manifold of accuracy and speed of real-time data transmission, combined with 5G technology of a GPS that helps to know the real-time location and increases the accuracy. Introducing new services in agriculture, construction, defense, aerial photography, marketing, delivery, rescue, and entertainment will expand the use of drones, thanks to high-definition cameras that record and transmit real-time video of a location while flying at a high altitude. Particularly in small-scale enterprises, delivering parcels by drones to regions with low productivity and short distances will be an important element in reducing operational costs. Using drones to rapidly deliver small packages for logistics, medicines, and food over short distances saves labor. However, it can also be offered as a solution to intensifying road traffic due to increasing urbanization.

Increasing environmental concerns and road traffic also increase the demand for air taxis. The widespread use of drone taxis will shorten travel times and reduce carbon emissions, cost, and traffic congestion. The spread of air ambulances will improve the rapid delivery of medical aid to areas in need and the use of drone ambulances in regions affected by natural disasters and congested areas. Drones equipped with thermal sensors, infrared, night vision cameras, and transmission devices, working as a powerful surveillance system, are one of the important tools in obtaining real-time information about the locations of missing persons, injured, and criminals in difficult and high terrain. In addition, drones facilitate tracking poachers, as they continuously monitor forest life and wildlife protection areas without approaching wild animals.

Drones, which provide a detailed view of large areas, spaces, and a particular subject, also intensify the interest of professionals such as commercial photographers, cartographers, and geologists and provide information about the environment at a low cost. Drones are the important helpers in ensuring the crowd's safety in demonstrations, marches, and other meetings held in public areas. Drones with high-definition cameras serve successfully in aerial image and video shooting at sports events.

Thermal sensors and gyroscopes, with their high maneuverability, small size, and power, increase the accuracy of regional and national weather forecasts by following the scientists' weather events in detail. It will be inevitable to use it more widely by developing in taking precautions and early warning works by giving an idea about the trajectory of large-scale weather events such as hurricanes.

Agriculture mechanization is important in developing agricultural practices for human nutrition and a sustainable, healthy food supply. The destruction caused by the increasing weight of the massive machines used for efficient and economical production in large areas is a reality. Drones have the potential to be an option in solving this problem. Particularly in agriculture 4.0, IoT applications, taking data quickly from large areas with unmanned aerial vehicles (UAV) and unmanned ground vehicles (UGV), swarm communicating with each other, mapping the land and field-specific applications of agricultural processes according to the data obtained accordingly, and product yield and quality. It has a sustainable solution potential that will increase.

9. Conclusions

Significant developments in drone technology in recent years have made the procurement cost of drones cheaper, and their use has become widespread for professional and civilian purposes. Although there are many different types of drones, it has become easy to find one that appeals to everyone. Drones are classified by fixed-wing, multi-rotor, or VTOL construction, size, weight, and power source. These features also determine the drone's range, flight time, and loading capacities. The transported load, cameras, sensors, and sensors can be hardware, parcels, drugs, and physical. In the control of the drone under load, certain frequencies are used for communication between the user controller and the drone. International coordination and standardization of this connection, called frequency spectrum in terms of flight and communication security, will fill a deficiency.

Future developments in drone technology include making drones smaller, lighter, more efficient, and cheaper. As such, drones will increasingly become available to the general public and be used for an ever-increasing scope of applications. Drones are expected to become more autonomous soon, and the ability to move in swarms will increase.

The main factor limiting the use of drones is the power supply, namely the flight time issue associated with the battery draining and needing to be recharged. On the other hand, the framework of the rules and sanctions that drone users must comply with regarding other citizens' privacy and personal rights while doing their targeted work are among the issues that should be clearly stated. In parallel with the developing technology, it is a fact that some features still considered a problem will be solved soon, and drones will become more visible in our daily lives.

Conflict of interest

The authors declare no conflict of interest.

DOI: http://dx.doi.org/10.5772/intechopen.1001987

Author details

Koç Mehmet Tuğrul
Faculty of Agriculture, Department of Biosystems Engineering, Eskişehir, Eskisehir Osmangazi University, Turkey

*Address all correspondence to: kmtugrul@gmail.com

References

[1] Lutkevich B. Drone (UAV) [Internet]. 2023. Available from: https://www.techtarget.com/iotagenda/definition/drone

[2] Floreano D, Wood RJ. Science, technology and the future of small autonomous drones (Review). Nature. 2015;**521**:460-466. DOI: 10.1038 / nature14542

[3] JOUAV. Different types of drones and uses (2023 Full Guide). 2022. Available from: https:// www. jouav.com/blog/drone-types.html#jouav-scrollspy-anchor-7

[4] Michael S. Drone anatomy 101. B&H The Professional's Source. 2017;**02**:210-211

[5] Akgüç B. FPV Camera Selection for RC/FPV planes and Drones. 2023. Available from: https://avdesodrone.com/rc-fpv-ucaklar-ve-dronlar-icin-fpv-kamera-secimi/

[6] Petropoulos G, Carlson TN, Wooster MJ, Islam S. A review of Ts / VI remote sensing-based methods for the retrieval of land surface energy fluxes and soil surface moisture. Progress in Physical Geography: Earth and Environment. 2009;**33**(2):224-250. DOI: 10.1177/0309133309338997

[7] Maes WH, Huete AR, Steppe K. Optimizing the processing of UAV-based thermal imagery. Remote Sensing. 2017;**9**(5):476. DOI: 10.3390/rs9050476

[8] Degondi P, Diez A, Na M. Bulk topological state in a new collectively dynamic model. 2021. Available from: https://www.researchgate.net/publication/348803228_Bulk_topological_ states_in_a_new_collective_dynamics_model /citations

[9] Wikipedia. Hyperspectral imaging. 2020. In Wikipedia. Available from: https://en.wikipedia.org/wiki/Hiperspektral_g%C3%B6r%C3%BCnt%C3%BCleme

[10] Corrigan F. Drone gyro stabilization, imu, and flight controllers explained. 2020. Available from: https://www.dronezon.com/learn-about-drones-quadcopters/three-and-six-axis-gyro-stabilized-drones/

[11] Dogan F. Drone Technologies. 2022. Available from: https://www.guvenlikyonetimi.com/dron-teknolojileri/

[12] Claesson A, Fredman D, Svensson L, et al. Unmanned aerial vehicles (drones) in out-of-hospital-cardiac-arrest. Scandinavian Journal of Trauma Emergency Medicine. 2016;**24**(1):124. Available from: https://www.ncbi. nlm.nih.gov/pubmed/27729058

[13] Scalea J, Restaino S, Scassero M, Blankenship G, Bartlett S, Wereley N. An initial investigation of unmanned aircraft systems (UAS) and real-time organ status measurement for transporting human organs. IEEE Journal of Translational Engineering Health Medicine. 2018;**6**:4000107. DOI: 10.1109/JTEHM.2018.2875704

[14] Daley S. Drone technology: What is a drone? 2023. Available from: https://builtin. com/drones

Chapter 2

Detection and Classification of Drones Using Radars, AI, and Full-Wave Electromagnetic CAD Tool

Ahmed N. Sayed, Omar M. Ramahi and George Shaker

Abstract

Detection and classification of drones have become crucial due to their potential usage in illicit activities. Radar systems can provide a promising solution to this needed task when combined with machine learning (ML) and artificial intelligence (AI) models. Radar datasets that contain drone information are needed to train AI models. Generating radar datasets that contain drone information is one of the most important challenges in this application as it is expensive and time-consuming. In addition, such datasets are limited to the radar used, the background environment, and drone types. In this chapter, full-wave electromagnetic (EM) and computer-aided design (CAD) tools are proposed for use to generate radar datasets that contain drone information. The proposed method overcomes this prevailing challenge in the field of radar detection and classification of drones. Furthermore, drones are widely classified using their range-Doppler information, which depends on their mechanical motions. The impact of the control systems of four different drones on their range-Doppler signatures is examined using a full-wave EM CAD tool. Finally, we demonstrate how we advance state-of-the-art literature on the detection and classification of drones utilizing radar systems, a mechanical control-based machine learning (MCML) algorithm is used to classify the four unmanned air vehicles (UAVs).

Keywords: radar detection, UAV classification, machine learning, numerical simulations, UAV control

1. Introduction

Unmanned air vehicles (UAVs), also known as drones, have become easily accessible worldwide in which they are possibly be used in many terrorist attacks and illegal activities [1–10]. This requires having systems to detect and classify drones at a distance to have time to take actions if needed. Radar-based systems are preferred in comparison with optical, acoustic, and RF-based systems due to their advantages in this application [11–16]. Radar systems work night and day, and in bad weather conditions, they can detect several drones at a time, track autonomous drones, and classify them when combined with ML models [11–16].

Classification of radar targets is widely achieved through the generation of range-Doppler images and micro-Dopler signatures [17–19]. Generating radar datasets that contain UAV information using real measurements costs a lot and wastes time. In addition, these datasets are limited to the used radars, UAV types, and the location and surroundings in which these measurements are made [20]. In this chapter, we propose to use a full-wave EM CAD tool, such as Ansys high-frequency structure simulator (HFSS) [21], to generate radar datasets the contain UAV information for the purpose of training ML algorithms [22–25]. Traditionally, full-wave EM CAD tools are used for designing and simulating high-frequency stationary electronic products. The proposed method using Ansys HFSS SBR+ solver [21–25] can be used to move drones and perform time-based full-wave analysis. For example, **Figure 1a–c** illustrate the DJI S900 hexacopter UAV, Ansys HFSS model of this hexacopter, and three-time stamps showing the rotation of its blades in Ansys HFSS, respectively.

2. Datasets generation

Six different UAVs are modeled using the full-wave EM tool, the six UAVs are a fixed-wing UAV, a helicopter UAV, 3 quadcopter UAVs, and a hexacopter UAV. They are modeled according to the Albatross drone [27, 28], the Black Eagle 50 drone [29], the DJI FPV drone [30], the MD4-1000 drone [31, 32], the Phantom 3 standard drone [33], and the DJI S900 drone [26], respectively. The specifications of these drones are shown in **Table 1**. A W-band frequency modulated continuous wave (FMCW) radar

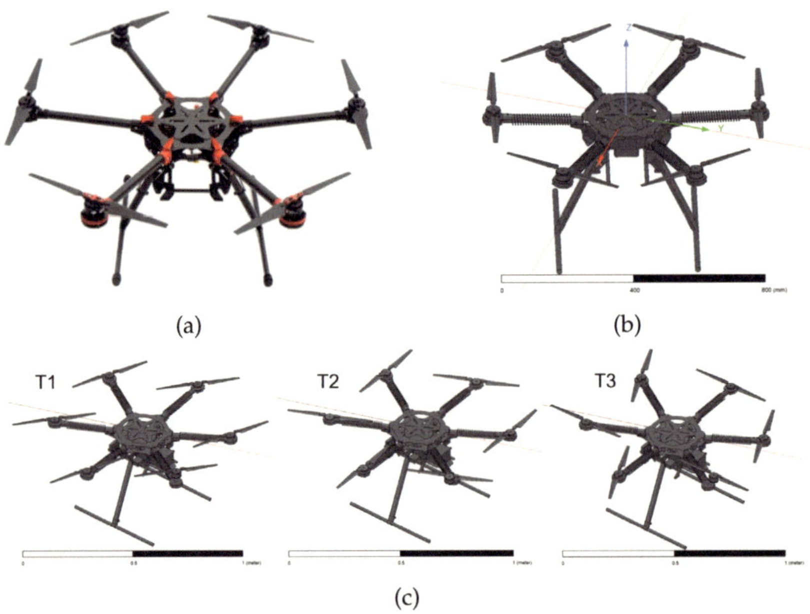

Figure 1.
(a) The DJI S900 UAV [26], (b) Ansys HFSS model for the DJI S900 UAV, and (c) three-time stamps for the modeled DJI S900 UAV showing its blades' rotation.

Type	UAV	Dimensions (m)
Fixed wing	Albatross	0.74 × 0.2 × 0.15
Helicopter	Black Eagle 50	2.65 × 0.56, blade 3.75
Quadcopter A	DJI FPV	0.178 × 0.232 × 0.127
Quadcopter B	MD4-1000	1.136 × 1.730 × 0.495
Quadcopter C	Phantom 3	Diagonal 0.35 and blade 0.24
Hexacopter	DJI S900	Diagonal 0.9 and arm 0.358

Table 1.
Dimensions of the six UAVs.

Quantity	Symbol	Value
Center frequency	f_0	77 GHz
Bandwidth	BW	300 MHz
Range resolution	ΔR	0.5 m
Velocity resolution	ΔV	0.4 m/s
Maximum range	R_{max}	60 m

Table 2.
Radar parameters.

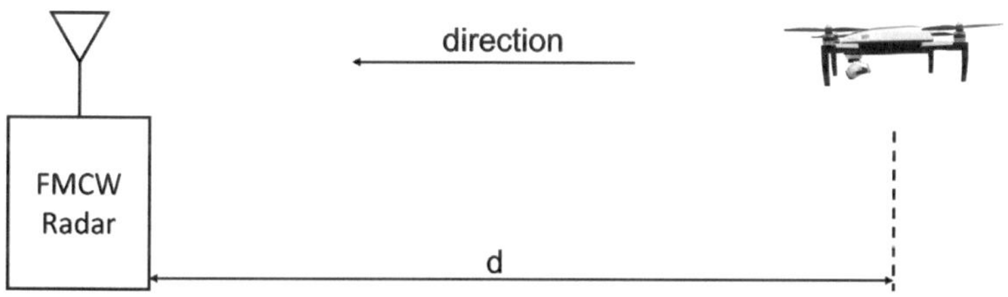

Figure 2.
Ansys HFSS test setup.

is used to generate the datasets required for this work for its high resolution. The radar parameters used in this work are shown in **Table 2**, these parameters are chosen for simplicity.

The six UAVs are modeled to pitch forward to the radar from a distance of 50 m/s with 5 m/s velocity, this distance is chosen to match the state-of-the-art literature using mmWave radars for this application [34–38]. **Figure 2** shows the Ansys HFSS simulation setup used to generate the required datasets for this work, while **Figure 3a–f** shows the range-Doppler maps obtained through Ansys HFSS for the Albatross, Black Eagle 50, DJI FPV, MD4 1000, DJI Phantom 3 standard, and DJI S900 UAVs, respectively.

A convolutional neural network (CNN) model is used to classify the six UAVs. The architecture of the CNN model showing the number, types, and dimensions of layers used in this work is illustrated in **Figure 4**. The CNN model is based on the DopplerNet CNN model [39], a max pooling layer is added to this model to decrease its complexity and avoid overfitting. The CNN model is applied to a dataset that

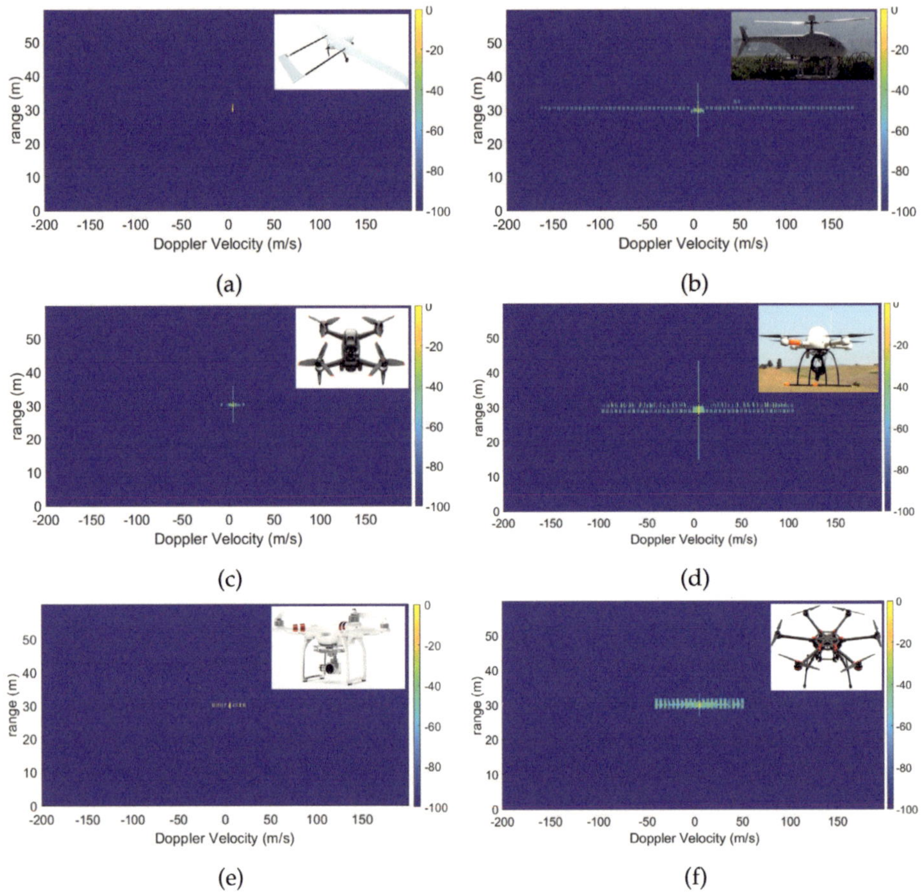

Figure 3.
Range-Doppler image for the (a) Albatross UAV, (b) Black Eagle 50 UAV, (c) DJI FPV UAV, (d) MD4 1000 UAV, (e) DJI Phantom 3 standard UAV, and (f) DJI S900 UAV.

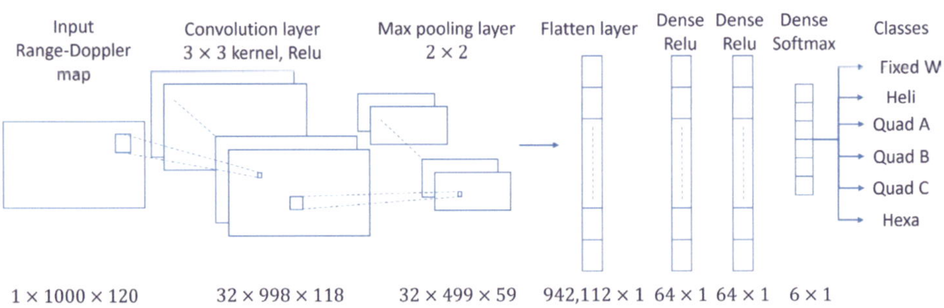

Figure 4.
The CNN model architecture.

contains a total of 1200 range-Doppler maps, 200 range-Doppler maps for each UAV. The classification result is found to exceed 97% as shown in **Figure 5**. The findings of this work demonstrate how accurate the proposed method can be used to generate radar datasets that contain UAV information and training ML models on them.

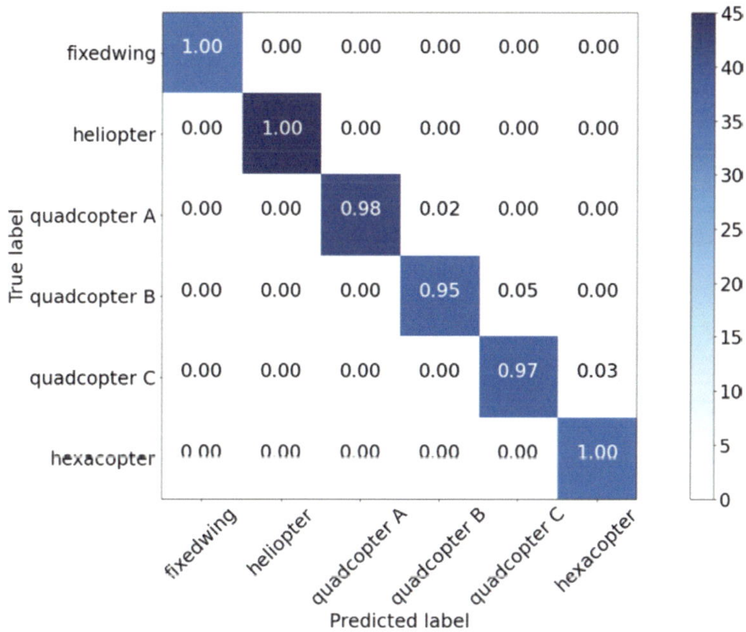

Figure 5.
The confusion matrix for the CNN model.

3. Mechanical control-based machine learning

UAVs have five basic motions, which are hovering, pitching, throttling, rolling, and yawing. To perform each motion, the speeds of a UAV's rotors have to be changed, which affects its Doppler signature [23]. In the state-of-the-art literature, ML models are trained on datasets that contain a UAV's hovering and pitching

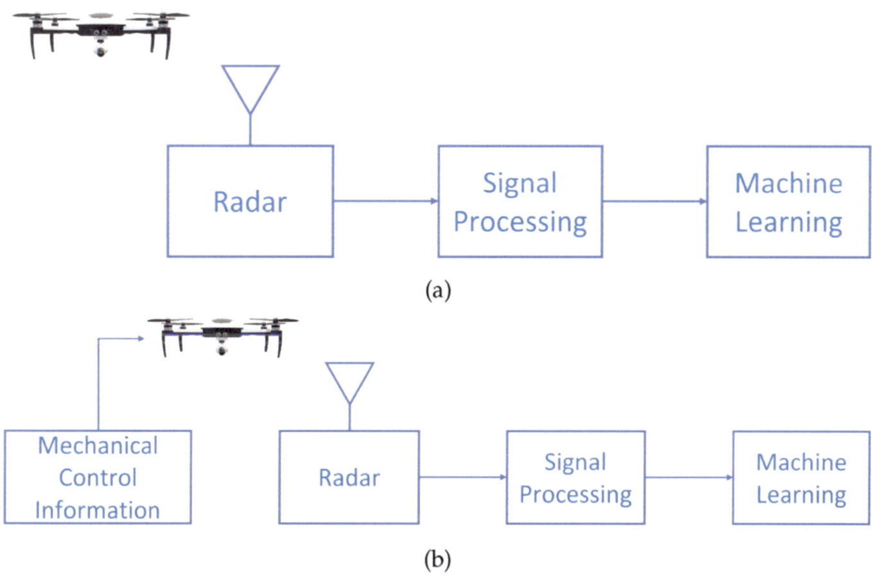

Figure 6.
(a) Traditional algorithm and (b) MCML algorithm.

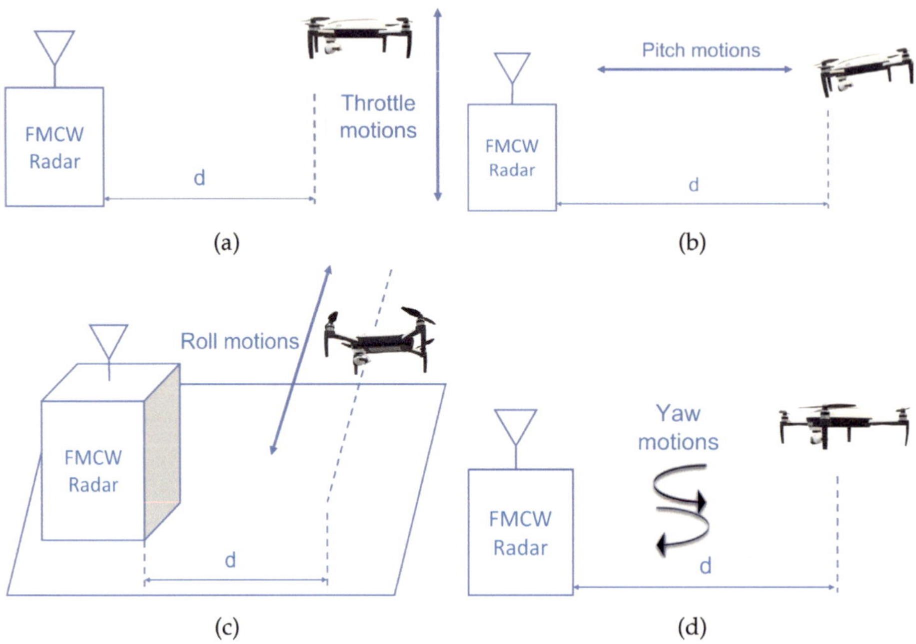

Figure 7.
Ansys HFSS simulation setups for different motions. (a) Throttle, (b) pitch, (c) roll, and (d) yaw.

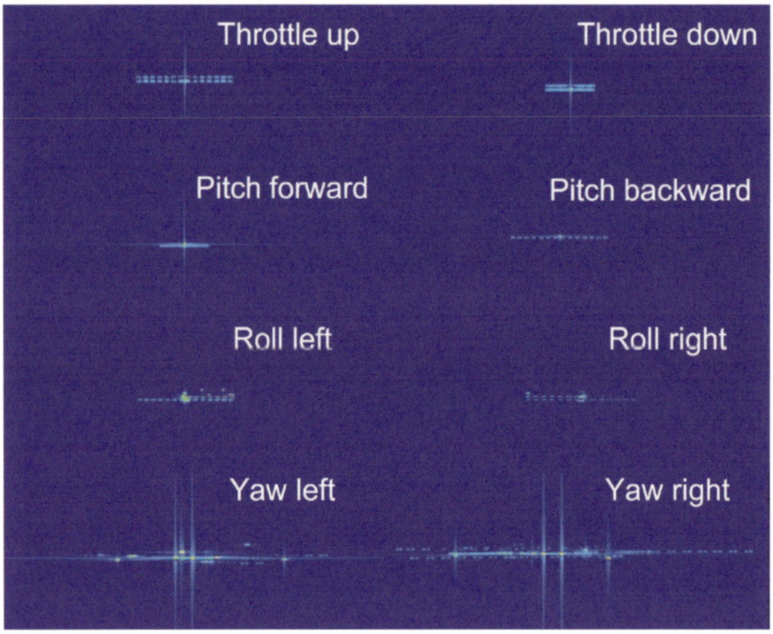

Figure 8.
Range-Doppler maps for the hexacopter UAV at different motions.

motions only, and they are tested on the same datasets [36, 40–50]. This leads to a degradation on ML accuracy if these trained models were tested using different datasets that contain other motions [23]. As radar signature of UAVs is highly dependent on the mechanical control information of them [23]. In this section, the impact of the mechanical control information of four UAVs on ML accuracy is investigated, and a mechanical control-based machine learning (MCML) algorithm is proposed to overcome this effect [23]. **Figure 6a** and **b** shows the traditional algorithm used for the state-of-the-art literature and the MCML algorithm, respectively [23].

To perform this investigation, the full-wave EM simulator is used to model the different motions of the four UAVs, a helicopter, a hexacopter, and two different quadcopters, **Figure 7a–d** show the throttle, pitch, roll, and yaw motions,

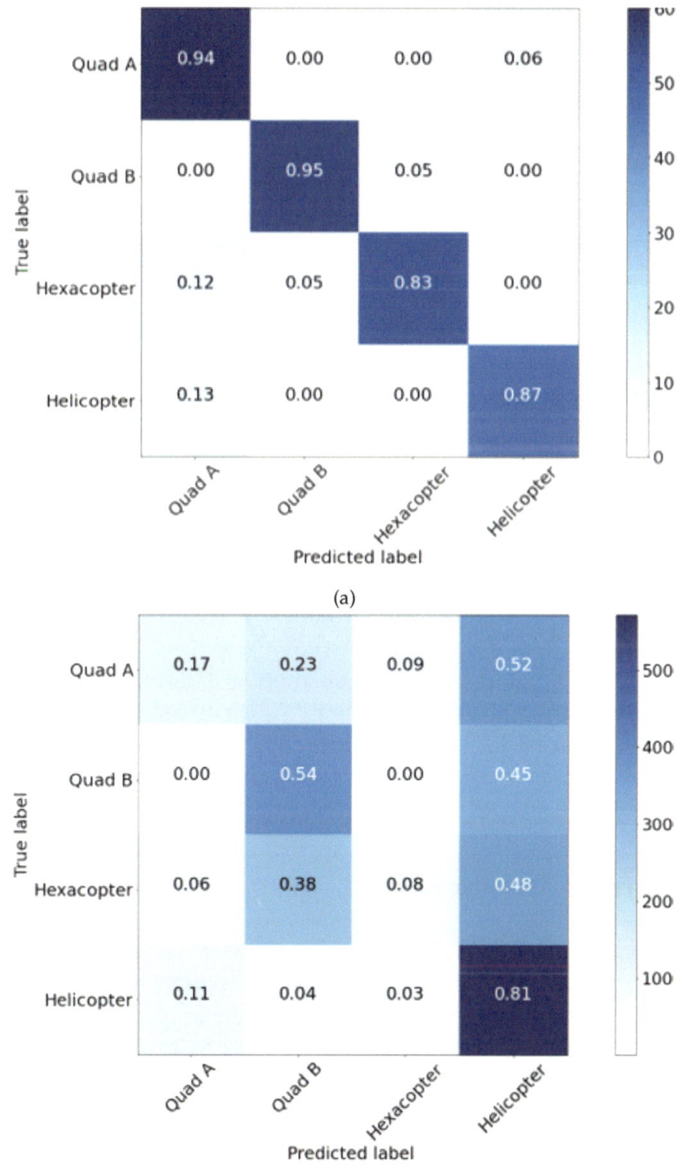

Figure 9.
The confusion matrix of the CNN model when it is tested on: (a) same dataset and (b) different dataset.

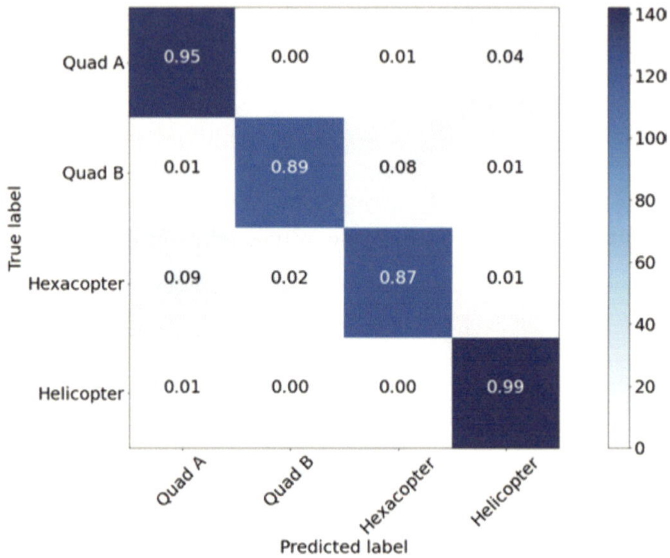

Figure 10.
The confusion matrix of the CNN model when applying the MCML method.

respectively [23]. As an example, the range-Doppler maps for the hexacopter UAV for the different motions are shown in **Figure 8**.

Two different datasets, containing the range-Doppler maps for the four UAVs, are generated: the first dataset contain the pitch and hover motions only to match the state-of-the-art literature, and the second dataset contain all the basic motions for the UAVs. The CNN model shown in **Figure 4** is trained on the first dataset and is tested on the same dataset as done in the state-of-the-art literature, and then it is tested on the second dataset that contains different motions to investigate the impact of the mechanical control information of the four UAVs. The result of this investigation is summarized in **Figure 9a** and **b**, which illustrates how the CNN model failed to classify the four UAVs when it was tested on another dataset that has different motions. Subsequently, the MCML algorithm [23] is applied, yielding an accuracy of 92.5% as shown in **Figure 10**. The MCML method gets over the loss in classification accuracy that occurs if the mechanical control information of UAVs is ignored.

4. Countering radar deception

Using numerical simulations to generate radar drones datasets facilitate the process for UAV and radar researchers to design and test appropriate radar systems for the detection and classification of drones. Furthermore, the proposed method can be used to decrease the effects of the attempts taken to deceive radar systems. Some of these several attempts that are not limited to modifying a UAV's body/blades, stealth-coating to reduce a UAV's radar cross section (RCS) area, equipping a standard UAV with explosives, and training birds to hide a UAV from radar detection. All these examples and more can be modeled using the proposed method to generate radar drones datasets and train ML models on them. For example, **Figures 11** and **12** and show the DJI S900 hexacpoter equipped by dynamite, and an octocopter is hidden by a group of birds, respectively. Radar signature of these cases can be generated to be studied, and in

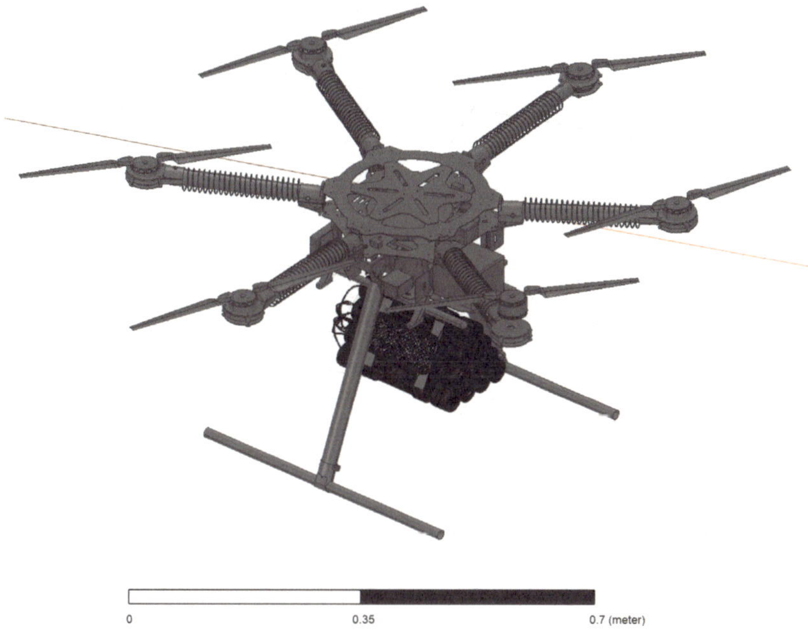

Figure 11.
Ansys HFSS model for the DJI S900 hexacopter UAV with dynamite attached.

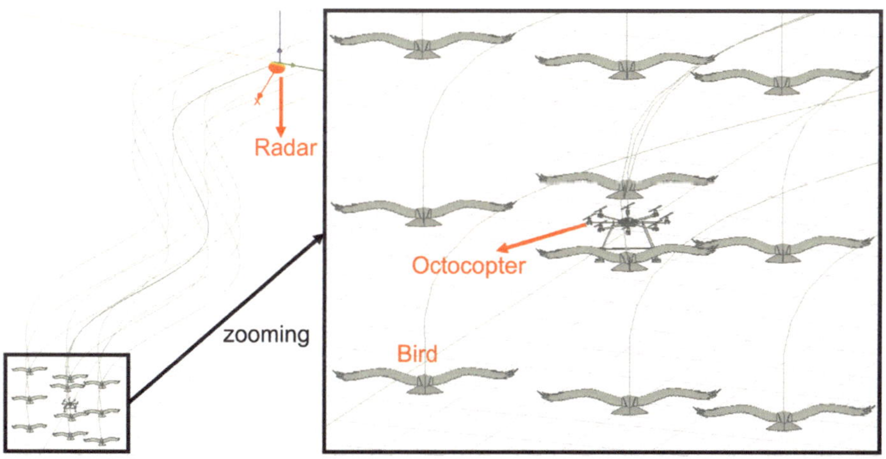

Figure 12.
Ansys HFSS model for a hidden octocopter by a group of trained birds.

addition, signal processing techniques and ML models can be developed to counter these attempts.

5. Conclusions

In this chapter, full-wave electromagnetic CAD tools are used to generate radar UAV's datasets to train machine learning (ML) models. The method provides accurate results for range and Doppler information of UAVs. The accuracy of a CNN model used in this work is found to exceed 97%. The proposed method presents a paradigm shift in how machine learning experts think about the application of radar classification of UAVs. The effect of the mechanical control information of UAVs on machine learning accuracy is explored using a full-wave electromagnetic CAD tool. The Doppler information is found to be highly dependent on mechanical control information of UAVs. A MCML method gets over the loss in classification accuracy that occurs if the mechanical control information of UAVs is ignored. The accuracy of the MCML algorithm is found to exceed 90% compared with the state-of-the-art literature in the application of radar detection and classification of UAVs.

Acknowledgements

This work was supported in part by NSERC. We would like to acknowledge CMC microsystems and Ansys for providing licenses for the CAD tools used throughout this work.

Conflict of interest

The authors declare no conflict of interest.

Abbreviations

ML	Machine learning
AI	Artificial intelligence
EM	Electromagnetic
CAD	Computer-aided design
FMCW	Frequency modulated continuous wave
MCML	Mechanical control-based machine learning
UAV	Unmanned air vehicle
HFSS	High-frequency structure simulator
BW	Bandwidth
CNN	Convolutional neural network
RCS	Radar cross section

Author details

Ahmed N. Sayed*, Omar M. Ramahi and George Shaker
Department of Electrical and Computer Engineering, University of Waterloo, Waterloo, Ontario, Canada

*Address all correspondence to: ansayed@uwaterloo.ca

References

[1] Heathrow airport: Drone sighting halts departures - bbc news. 2022. [Online]. Available from: https://www.bbc.com/news/uk-46803713

[2] Drone sightings keep closing london's airports - bloomberg. 2022. [Online]. Available from: https://www.bloomberg.com/news/articles/2019-01-09/drone-sightings-keep-closing-london-s-airports

[3] Shrapnel injures 12 at saudi abha airport as drone intercepted - reuters. 2022. [Online]. Available from: https://www.reuters.com/world/middle-east/saudi-led-coalition-says-destroyed-drone-launched-towards-abha-airport-4-injured-2022-02-10/

[4] Iraq assassination attempt signals new era of non-state drone attacks - middle east eye. 2022. [Online]. Available from: https://www.middleeasteye.net/opinion/iraq-assassination-new-era-drone-non-state-actors

[5] Turkey closes diyarbakir airport for month after drone attack 'by pkk' - world - the times. 2022. [Online]. Available from: https://www.thetimes.co.uk/article/turkey-closes-diyarbakir-airport-for-month-after-drone-attack-by-pkk-gmcj8skfz

[6] Drone attack in abu dhabi kills 3 wounds 6 - cbc news. 2022. [Online]. Available from: https://www.cbc.ca/news/world/abu-dhabi-drone-attack-1.6317555

[7] Flights diverted at east midlands airport after drone sightings - air transport - the guardian. 2022. [Online]. Available from: https://www.theguardian.com/world/2022/jun/11/flights-diverted-at-east-midlands-airport-after-drone-sightings

[8] Rcmp charge two from b.c. after drone used to smuggle drugs into prison - ctv news. 2022. [Online]. Available from: https://winnipeg.ctvnews.ca/drone-used-to-smuggle-drugs-into-manitoba-prison-rcmp-charge-two-men-from-b-c-1.5980061

[9] Small drones are giving ukraine an unprecedented edge - wired. 2022. [Online]. Available from: https://www.wired.com/story/drones-russia-ukraine-war/

[10] Ukraine: How drones are changing the way of war - science - in-depth reporting on science andtechnology - dw. 2022. [Online]. Available from: https://www.dw.com/en/ukraine-how-drones-are-changing-the-way-of-war/a-61681013

[11] Lykou G, Moustakas D, Gritzalis D. Defending airports from uas: A survey on cyber- attacks and counter-drone sensing technologies. Sensors. Jun 2020; **20**(12):1-35

[12] Park S, Kim HT, Lee S, Joo H, Kim H. Survey on anti-drone systems: Components, designs, and challenges. IEEE Access. Mar 2021;**9**42 635-42 659

[13] Samaras S, Diamantidou E, Ataloglou D, Sakellariou N, Vafeiadis A, Magoulianitis V, et al. Deep learning on multi sensor data for counter uav applications—a systematic review. Sensors. Nov 2019;**19**22:4837

[14] Flórez J, Ortega A, Betancourt A, García M, Bedoya, Botero JS, A review of algorithms, methods, and techniques for detecting uavs and uas using audio, radiofrequency, and video applications. TecnoLógicas. May 2020;**23**(48): 269-285

[15] Sedunov A, Haddad D, Salloum H, Sutin A, Sedunov N, Yakubovskiy A. Stevens drone detection acoustic system and experiments in acoustics uav tracking. In: 2019 IEEE International Symposium on Technologies for Homeland Security (HST). Woburn, MA, USA: Institute of Electrical and Electronics Engineers Inc.; Nov 2019

[16] Balachandran V, Sarath S. A novel approach to detect unmanned aerial vehicle using pix2pix generative adversarial network. In: 2022 Second International Conference on Artificial Intelligence and Smart Energy (ICAIS). Coimbatore, India: Institute of Electrical and Electronics Engineers Inc.; 2022, pp. 1368-1373

[17] Kang KB, Choi JH, Cho BL, Lee JS, Kim KT. Analysis of micro-doppler signatures of small uavs based on doppler spectrum. IEEE Transactions on Aerospace and Electronic Systems. Oct 2021;**57**(5):3252-3267

[18] Sun H, Oh BS, Guo X, Lin Z. Improving the doppler resolution of ground-based surveillance radar for drone detection. IEEE Transactions on Aerospace and Electronic Systems. Dec 2019;**55**(6):3667-3673

[19] Oh BS, Lin Z. Extraction of global and local micro-doppler signature features from fmcw radar returns for uav detection. IEEE Transactions on Aerospace and Electronic Systems. Apr 2021;**57**(2):1351-1360

[20] Hanif A, Muaz M, Hasan A, Adeel M. Micro-doppler based target recognition with radars: A review. IEEE Sensors Journal. Feb 2022;**22**(4): 2948-2961

[21] Ansys inc. 2022. [Online]. Available from: https://www.ansys.com/products/electronics/ansys-hfss

[22] Sayed AN, Riad MMYR, Ramahi OM, Shaker G. A methodology for uav classification using machine learning and full-wave electromagnetic simulations. In: 2022 International Telecommunications Conference (ITC-Egypt). Alexandria, Egypt; Aug 2022. pp. 1-2

[23] Sayed AN, Ramahi OM, Shaker G. Machine learning for uav classification employing mechanical control information. IEEE Transactions on Aerospace and Electronic Systems. May 2023:1-14. DOI: 10.1109/TAES.2023.3272303

[24] Sayed AN, Abedi H, Ramahi OM, Shaker G. On the impact of an antenna field of view on the classification of uavs. In: 2023 International Workshop on Antenna Technology (iWAT). Aalborg, Denmark. May 2023. pp. 1-2

[25] Sayed AN, Ramahi OM, Shaker G. Uav classification utilizing radar digital twins. In: 2023 IEEE International Symposium on Antennas and Propagation and USNC-URSI Radio Science Meeting (USNC-URSI). Portland, OR, USA. 2023. pp. 741-742

[26] Spreading wings s900 - dji. 2022. [Online]. Available from: https://www.dji.com/ca/spreading-wings-s900

[27] Albatross uav: Long range drone. 2022. [Online]. Available from: https://www.appliedaeronautics.com/albatross-uav/

[28] Fixed-wing drone, commercial long range uav, applied aeronautics. 2022. [Online]. Available from: https://www.unmannedsystemstechnology.com/company/applied-aeronautics/

[29] Unmanned helicopter, small rotary, tactical uav, isr drones, ruav. 2022.

[Online]. Available from: https://www.unmannedsystemstechnology.com/company/steadicopter/

[30] Dji fpv - specs - dji. 2022. [Online]. Available from: https://www.dji.com/ca/dji-fpv/specs

[31] Torres-Sánchez FL, Pez-Granados D, Castro, Barragán AIPA. Configuration and specifications of an unmanned aerial vehicle (uav) for early site specific weed management. PLoS ONE. Mar 2013;**8**(3): 58210

[32] Microdrones md4-1000 full specifications & reviews. 2022. [Online]. Available from: https://productz.com/en/microdrones-md4-1000/p/y5RW

[33] Phantom 3 standard - dji. 2023. [Online]. Available from: https://www.dji.com/ca/phantom-3-standard

[34] Semkin V, Haarla J, Pairon T, Slezak C, Rangan S, Viikari V, et al. Analyzing radar cross section signatures of diverse drone models at mmwave frequencies. IEEE Access. Mar 2020;**8**:48 958-48 969

[35] Karlsson A, Jansson M, Hämäläinen M. Model-aided drone classification using convolutional neural networks. In: 2022 IEEE Radar Conference (RadarConf22). New York City, NY, USA. 2022. pp. 1-6

[36] Rai PK, Idsoe H, Yakkati RR, Kumar A, Khan MZA, Yalavarthy PK, Cenkeramaddi LR. Localization and activity classification of unmanned aerial vehicle using mmwave fmcw radars. IEEE Sensors Journal. Jul 2021;**21**(14):16 043-16 053

[37] Fu R, Al-Absi MA, Kim KH, Lee YS, Al-Absi AA, Lee HJ. Deep learning-based drone classification using radar cross section signatures at mmwave frequencies. IEEE Access. Sep 2021;**9**:161 431-161 444

[38] Caris M, Johannes W, Sieger S, Port V, Stanko S. Detection of small uas with w-band radar. In: 2017 18th International Radar Symposium (IRS). Prague, Czech Republic, 2017. pp. 1-6

[39] Roldan I, Del-Blanco CR, De Quevedo ÁD, Urzaiz FI, Menoyo JG, Ĺopez AA, et al. Dopplernet: A convolutional neural network for recognising targets in real scenarios using a persistent range-doppler radar. IET Radar, Sonar and Navigation. Apr 2020;**14**(4):593-600

[40] Park J, Park J-S, Park S-O. Small drone classification with light cnn and new micro-doppler signature extraction method based on a-spc technique. Arxiv. Sep 2020

[41] Kim BK, Kang HS, Park SO. Drone classification using convolutional neural networks with merged doppler images. IEEE Geoscience and Remote Sensing Letters. Jan 2017;**14**(1):38-42

[42] Zhang P, Yang L, Chen G, Li G. Classification of drones based on micro-doppler signatures with dual band radar sensors. In: 2017 Progress in Electromagnetics Research Symposium - Fall (PIERS - FALL). Singapore: Electromagnetics Academy; Nov 2017. pp. 638-643

[43] Zhang P, Li G, Huo C, Yin H. Exploitation of multipath micro-doppler signatures for drone classification. IET Radar, Sonar & Navigation. Apr 2020;**14** (4):586-592

[44] Zhang W, Li G, Baker C. Radar recognition of multiple micro-drones based on their micro-doppler signatures

via dictionary learning. IET Radar, Sonar & Navigation. Sep 2020;**14**(9): 1310-1318

[45] Rahman S, Robertson D. Time-frequency analysis of millimeter-wave radar micro-doppler data from small uavs. In: 2017 Sensor Signal Processing for Defence Conference (SSPD). London, UK: Institute of Electrical and Electronics Engineers Inc.; Dec 2017. pp. 1-5

[46] Li CJ, Ling H. An investigation on the radar signatures of small consumer drones. IEEE Antennas and Wireless Propagation Letters. Jul 2017;**16**:649-652

[47] Kumawat HC, Chakraborty M, Raj AAB, Dhavale SV. Diat-"μsat: Small aerial targets' micro-doppler signatures and their classification using cnn. IEEE Geoscience and Remote Sensing Letters. Aug 2021;**19**:1-5

[48] Patel JS, Al-Ameri C, Fioranelli F, Anderson D. Multi-time frequency analysis and classification of a micro-drone carrying payloads using multistatic radar. The Journal of Engineering. Oct 2019;**2019**(20): 7047-7051

[49] Ritchie M, Fioranelli F, Borrion H, Griffiths H. Multistatic micro-doppler radar feature extraction for classification of unloaded/loaded micro-drones. IET Radar, Sonar and Navigation. Jan 2017;**11** (1):116-124

[50] Mendis GJ, Randeny T, Wei J, Madanayake A. Deep learning based doppler radar for micro uas detection and classification. In: MILCOM 2016 - 2016 IEEE Military Communications Conference. Baltimore, MD, USA: Institute of Electrical and Electronics Engineers Inc.; Dec 2016. pp. 924-929

Chapter 3

Defining and Implementing Autonomy in Multi-Drone Systems

Ogbonnaya Anicho

Abstract

Defining autonomy or autonomous capability from the context of a single drone is often the default position. However, future drone applications will deploy as multi-drone systems comprising multiple drones. Such systems will carry out specific and complex tasks cooperatively in various use scenarios. It is imperative to understand how autonomy for these multi-drone systems could be better defined for design, regulatory and operational purposes. The chapter proposes a framework for defining and evaluating autonomy for multi-drone systems by segregating the system into hierarchies and layers. In the work, a typical multi-drone system is segregated into three (3) layers consisting of; Single Vehicle Control (Layer 1), Multi-Vehicle Control (Layer 2) and Global Mission Control (Layer 3). This framework could be beneficial to designers, regulators and standardisation efforts. Currently, some progress is in motion to find a consensus on the definition and ramifications of autonomy levels and human-autonomy interactions. This chapter contributes to the ongoing efforts by proposing a framework that addresses autonomy or autonomous capability for multi-drone systems.

Keywords: drone, UAV, autonomy, autonomous drones, multi-drone, UAS operations

1. Introduction

A drone which is also known as an Unmanned Aerial Vehicle (UAV) is a powered aerial platform or vehicle that does not have a human pilot physically on board. It can be designed to fly autonomously or remotely piloted hence also called a remotely piloted aircraft (RPA). According to the UK Civil Aviation Authority (CAA) and the European Union Aviation Safety Agency (EASA), an Unmanned Aircraft (UA) is 'any aircraft operating or designed to be operated autonomously or to be piloted remotely without a pilot on board' [1, 2]. The UK CAA & EASA also distinguish between autonomous and automatic operations. In automatic or automated operations, the drone follows pre-programmed instructions with the Remote Pilot (RP) still able to intervene in the flight [1, 2]. The term unmanned aircraft system (UAS) [2] is used when all other aspects of the system are considered including those on the ground. A typical UAS will include the UAV, ground control station (GCS), payload, control and data link and all other supporting equipment [3]. It is essential to lay out these key definitions to enhance our understanding and considerations for autonomy or autonomous capabilities in drones. Drones as shown in **Figures 1** and **2**, are mainly multirotor or fixed-wing (or some form of hybrid) with consequential technical and operational differences.

UAV autonomy is often defined from the context of a single drone and its capabilities to make decisions with minimal or no human input. The use of single remotely piloted drones within visual line of sight (VLOS) is commonplace and implemented across several use cases. However, the application of fully autonomous multi-drone systems is still futuristic. Multi-drone systems in the context of this chapter refer to the use of multiple drone (multi-drone) platforms in a coordinated fashion to carry out a specific or diverse array of tasks. It is important to distinguish the context and definition of autonomous multi-drone systems from swarming or drone swarms and the implications of the concept of systems.

Figure 1.
Multirotor drone [1].

Figure 2.
Fixed-wing drone [4].

Drones are fast becoming mainstream and increasingly deployed within our industrial, social and digital domains to solve various challenges, for example, Agriculture, traffic management, logistics & last-mile delivery, power line inspection, connectivity and a host of other areas [4, 5]. According to GSMA, it is estimated that drones can boost Gross Domestic Product (GDP) by 1.6% in the UK alone by 2030 [6]. Globally, this trend is also noticeable and growing across industries and applications. Drone startups and technology companies are driving innovation and solutions with drones globally. On the regulatory front, efforts are ongoing to provide adequate oversight and support to find the right balance between safety and security on one hand and innovation on the other. The regulatory aspect of the drone business will be crucial in determining its short- and long-term viability, especially from a commercial standpoint. One of the key challenges is safely integrating them with the rest of the manned airspace. Currently, most drones operate outside controlled or restricted airspace minimising incidences with other airspace users [5]. This is in addition to the significant challenge of regulating growing drone traffic with aeroplanes, helicopters and other aerial systems.

The main contribution of this paper is to propose a framework or model to define the autonomy of multi-drone systems which are still in development or futuristic at this time. This framework will help understand the challenges of designing, implementing and regulating the operations of such autonomous multi-drone use cases. Regulators have been cautious in approving the use of fully autonomous drone operations and have constantly requested input from stakeholders. This chapter is expected to contribute some ideas to the ongoing conversation on the parameters for implementing safe and autonomous drone systems. Section 1, introduces some key definitions and context of the chapter. In Section 2, the concept of drone autonomy/ autonomous operations is evaluated. BVLOS operations and the concept of autonomy were addressed in Section 4. In Section 5, the proposed framework detailing hierarchy and layers of autonomy for both single and multi-drone systems was introduced. Finally, the conclusions drawn by the author are captured in Section 6.

2. Defining autonomy or autonomous operations for drones

Generally, autonomy or autonomous capability is defined within the context of decision-making or self-governance in a system. According to the Aerospace Technology Institute (ATI), autonomous systems fundamentally can decide by themselves how to achieve the objectives of a mission without any human intervention [7]. These systems are also capable of learning and adapting to the changing state of the operating environment. Autonomy, however, is defined in levels and may depend on the design, functions and specifics of the mission or system [8]. Autonomy more broadly can be viewed as a spectrum of capabilities ranging from zero autonomy to full autonomy. The Pilot Authority and Control of Tasks (PACT) assigns levels of authority, from level 0 (full human pilot authority) to level 5 (full system autonomy), also applied to the automotive industry for autonomous vehicles (see **Figure 3**) [7]. Another general but useful model for describing levels of autonomy in unmanned systems is the Autonomous Levels For Unmanned Systems (ALFUS) [9].

The EASA in one of its technical reports provided some insight on autonomy levels and guidelines for human-autonomy interactions. According to the EASA, the concept of autonomy, its levels and human-autonomous system interactions are not settled and remain actively discussed in different domains (including aviation) as no

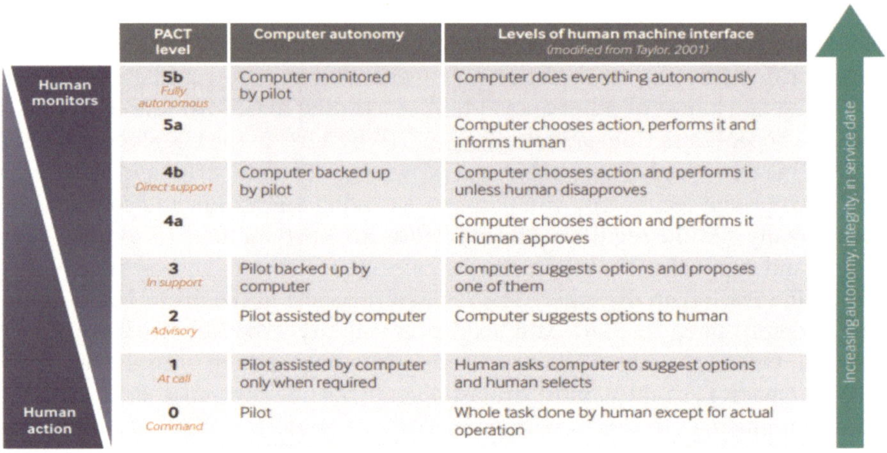

Figure 3.
Pilot authority and control of tasks (PACT) [7].

common understanding of these terms currently exists [2]. Since these concepts are still fluid in a sense, it becomes a huge challenge for the UAS regulatory environment as these concepts remain largely unsettled. This chapter attempts to provide a framework that may address some of the current challenges around a common understanding of concepts across jurisdictions. A framework that will have universal appeal must be uncomplicated and effective in addressing the concerns of regulators across jurisdictions.

3. Autonomy in multi-drone systems

Multi-Drone Systems in the context of this chapter refer to systems or implementations where multiple drones work together to complete a specific task. In these implementations, a single drone cannot deploy singularly regardless of sophistication. It is also technically not a swarm since a swarm by definition constitutes simple entities or agents interacting locally with their environments and jointly achieving an emergent advanced global behaviour [10] which far outweighs what any agent could have achieved as a single entity. The fascinating murmuration of a flock of starlings is a good example of how simple local actions can have an incredible global output. However, in multi-drone systems, individual drones are not expected to be simple and oblivious to the global mission. The design of multi-drone systems requires individual drones to have elevated levels of autonomous capabilities. Some of the tasks or areas where multiple drones can be implemented are;

- Search and Rescue Missions
- Network Area Coverage and Connectivity
- Space Specific Explorations
- Exploration and Extraction in Hostile Environments

- Sensitive Material Handling and Transportation
- Coordinated Wildfire Containment

Some of the use cases cited above are currently being resourced by single drones working solo. For instance, the NASA ingenuity helicopter deployed for Mars exploration is a single drone (see **Figure 4**) [11]. Upgrading this single drone to a fully autonomous multi-drone system will entail having about 2 or more of these helicopters working in a coordinated and autonomous mode on the surface of Mars. This is also the case with the firefighting support drone (see **Figure 5**), used by the Spanish Emergency Military Unit (UME) in emergencies like wildfires and natural disaster responses [12]. These firefighting drones at the moment contribute mainly towards providing timely information to commanders on the ground about the progress of the

Figure 4.
Ingenuity - NASA's Mars helicopter [11].

Figure 5.
Spanish UME firefighting drone [12].

fires whilst reducing the risks to those on the ground. However, in a fully autonomous multi-drone firefighting system, drones can coordinate and contain or extinguish fires with minimal or no human intervention using artificial intelligence and machine learning-based capabilities. This is the desired trajectory for multi-drone systems, however, this has to be done with the highest safety case.

Whilst the autonomous capabilities for single drones can be technically easier to define and evaluate, the question is can these same considerations be extended to a multiple drone system? As shown in **Figures 6** and 7, most operational or commercial drone applications are mainly single drone operational models. For example, Zipline uses single drone operations to deliver its autonomous aerial logistics business. It provides on-demand blood delivery, vaccines and many essential medical products using its fixed-wing drones [13]. The key consideration here is that the logistics use case being delivered by Zipline does not meet the definition of a multi-drone system.

Figure 6.
Zipline's drone logistics use case [13].

Figure 7.
Alphabet's drone delivery use case [14].

The same applies to Aplhabet's drone subsidiary which employs a single drone operational mode for its drone delivery business.

4. Beyond visual line of sight (BVLOS) and autonomy

Aviation authorities are careful to ensure that safety is of the highest consideration in the operations of any manned or unmanned flight. This becomes even more critical as the weight of the vehicle increases or flies over populated areas. In the civil UAV domain, the mode of operation mostly supported by regulation is one in which the UAV is within the pilot's visual line of sight (VLOS), (between 500 m and 2 km) [15]. The International Civil Aviation Organisation (ICAO) insists that VLOS means a straight line along which the remote pilot or UA observer has a clear view of the UA [16]. This mode of operation is cautiously considered to minimise safety and security hazards that may result from flying beyond visual sight. However, this approach imposes significant challenges on innovation and prevents and imposes limitations on scaling the operations. The conservative and cautious regulatory position on keeping VLOS as the default mode is understandable but stifles innovation in the civil UAV domain. Another option for UAS operations is the extended visual line of sight (EVLOS), where the pilot's view is extended by using other human observers positioned within the visual line of sight of the drone. The EASA guidance document defines BVLOS operations broadly as any UAS operations that are not conducted under VLOS conditions which specifically require the remote pilot to maintain continuous unaided visual contact with the unmanned aircraft [2] (see **Figure 8**).

BVLOS, which involves the UAV being able to continue operations beyond the pilot's field of view or observation, is highly restricted for civil UAS operations in most jurisdictions globally [15]. To operate safely and fly autonomously, drones will need to be BVLOS certified with reliable and secure methods for control [4]. BVLOS operations, however, have been in use by advanced military and defence establishments for a long time now. In essence, the issue is not just about the technological capabilities required for BVLOS but its management, integration and operational impact on the

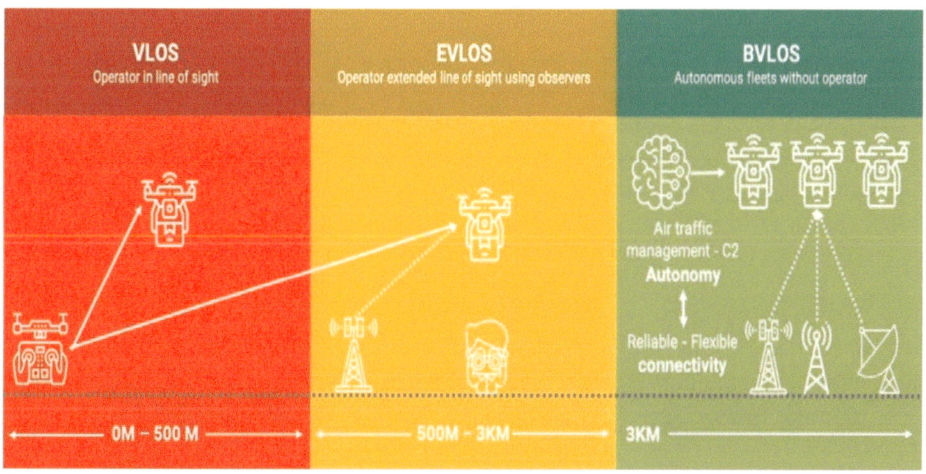

Figure 8.
Illustrating VLOS, EVLOS & BVLOS [17].

safety and security of the civil aviation environment. That is why this chapter is more focused on the autonomy models or frameworks that will aid the management, operations and regulation of UAS within the civil aviation domain.

Currently, a jurisdiction's BVLOS approval framework can be used to gauge appetite and readiness for increased levels of autonomous UAS operations. The current Federal Aviation Authority (FAA) Part 107 rules for commercial drone operations prohibit flying BVLOS [18], though special waivers and approvals are granted on a case-by-case basis to companies testing or trialling use cases. BVLOS operations can unlock the potential of the drone industry and specifically the multi-drone use cases. It will enable the capability to use connectivity, cloud infrastructure and other enabling technologies to scale multi-drone applications. However, one crucial requirement for BVLOS operations is a highly reliable C2 link which is imperative to fulfilling the BVLOS safety case. In terms of defining autonomy especially in multi-drone operations, is BVLOS capability the ultimate expression of autonomy? These questions are important to guide and inform regulatory oversight of multi-drone implementations and use cases.

5. Defining and segregating autonomy into hierarchies/layers

One approach to addressing the autonomy challenge is to segregate autonomy in drones and multi-drone systems into hierarchies/layers. These layers will have standard definitions and protocols to guide technology development and regulatory oversight. The author proposes two distinct layers for single drone autonomy models which will consist of Vehicle/Platform Control (Layer 1) and Mission Control (Layer 2), see **Figure 9**. As shown in **Figure 9**, multi-drone systems on the other hand will have three (3) layers consisting of Single Vehicle/Platform Control (Layer 1), Multi-Vehicle/Platform Control (Layer 2) and Mission Control (Layer 3). These layers or hierarchies are not necessarily physical but logical to provide both conceptual and functional ways to manage the complexities in design, technology standards, regulation and operations. It is important to ensure that all aspects of the framework and supported protocols are open which will drive innovation and lower barriers to entry.

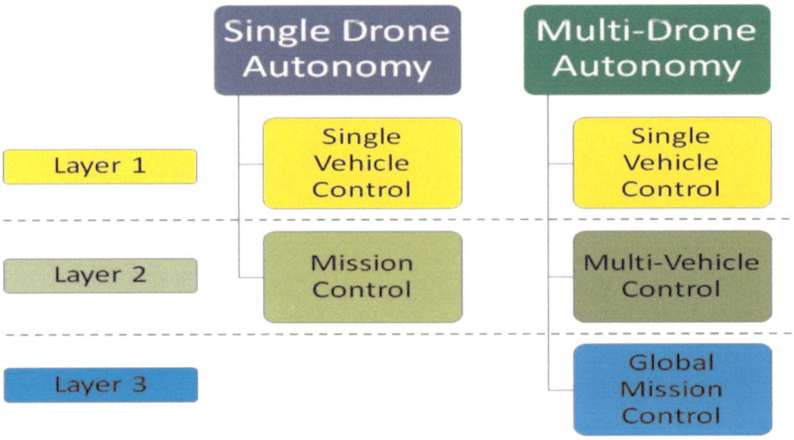

Figure 9.
Autonomy hierarchies/layers for multi-drone systems.

5.1 Hierarchy and layers for single drone autonomy

The autonomy hierarchy and layers for a single drone are outlined below and capture a conceptual model that is consistent with current implementations.

- **Single Vehicle Control - Layer 1:** This layer occupies the highest position in the single drone autonomy hierarchy or model. In practical terms, this layer ensures that the drone remains airborne operating safely and able to return/land safely even in the event of operational challenges without human pilot intervention. In a typical single drone setup, this layer includes the Flight Control System (FCS) and any other hardware/software needed to ensure the vehicle can fly safely and securely without collision or stalling from one set point to another.

- **Mission Control - Layer 2:** In a single drone system, the mission control layer is tasked with making decisions and fulfilling the objective of the mission as defined. It is the mission managing layer and ensures that the drone can autonomously handle the mission. In a typical single drone setup, the mission control layer will largely need some sort of additional computing hardware to process mission control tasks (e.g., signals from different sensors). It is currently common to find companion computers supporting the mission control tasks added as extra hardware separate from the FCS. In more advanced systems, integrated units can handle both vehicle and mission control tasks, however, the hierarchy (as shown in **Figure 9**) should prioritise the vehicle control layer with the authority to override mission control in critical scenarios. It is therefore necessary for designers of autonomous drones to ensure that this hierarchy model is clearly defined at the fundamental level of the drone design philosophy. There is also the design consideration to be made in terms of the level of autonomy each layer should be tuned to. It is also possible to experiment with different permutations of autonomy levels (PACT levels) to find the optimal combination. However, the ideal goal or objective is to have both the vehicle and mission control layers operating at full autonomy levels. Notably, the single vehicle autonomy model does not have a third layer.

5.2 Hierarchy and layers for multi-drone drone autonomy

The multi-drone autonomy model has 3 layers, unlike the single drone case which has only 2. It is important to emphasise that multi-drone systems are not just multiple drones flying autonomously. Multi-drones in this context are fully autonomous drones flying in a coordinated fashion to achieve a specific goal which cannot be done by any single one of them. In such a scenario, the operational environment is more complex than the single drone scenario. In these types of multi-drone systems, the layers of autonomy are explained further below;

- **Single Vehicle Control - Layer 1:** A multi-drone system is made up of single drones with autonomous capabilities. In a multi-drone setup, the single vehicle control layer occupies the same hierarchy and serves the same functions outlined in the single drone model. It is important to establish that the autonomy model proposed in this work elevates the hierarchy of the single drones that make up the autonomous multi-drone system. These single drones are expected to have elevated levels of intelligence and decision-making capabilities for the multi-drone system to fulfil its mission, which is somewhat contrary to the philosophy of swarming or swarm intelligence-based models. In essence, the vehicle control layer is the same for both single and multi-drone systems. As a practical matter, a single drone should have sufficient autonomy to make decisions that may override the global mission objective if prevailing conditions threaten its safety or security. Indeed, it is due to issues of this nature that the author proposes a model of evaluating autonomy for multi-drone systems.

- **Multi-Vehicle Control - Layer 2:** In the multi-drone system autonomy framework, the second layer is the multi-vehicle control layer tasked with coordinating the multi-vehicle decision process. This layer introduces the complexity inherent in multi-drone systems. For the first time, the issue of coordination and information exchange comes into focus. In some sense, this layer is difficult to simulate or conceptualise. Centralised control is not supported as it defeats the concept of autonomy. Participating drones retain a subset of the overall control structure. It is challenging to define autonomy at this layer as a cumulative concept. The author would rather define autonomy at this layer as a distributed and dynamic concept or capability that fluctuates based on what happens in Layer 1. The structure of this layer must be robust and flexible to adapt as local states in Layer 1 keep evolving. The ability of the entire system to manage and learn from these changes is key to survival and fulfilling mission goals. Depending on the use case or mission, the constituent drones in the system have to make decisions relevant to their local environments whilst tracking impact on the entire stability and posture of the fleet. How to functionally define and implement the autonomy existing at this layer makes the multi-drone system autonomy a challenging concept for design and regulation. This layer could be viewed as a bridge linking the single vehicle control layer to the global mission control layer.

- **Global Mission Control - Layer 3:** This is the layer that handles the actual mission or the objective of the multi-drone system. It ensures that the global mission i.e. the overall objective of the multi-drone application is achieved. For instance, an autonomous multi-drone system that is tasked with identifying and extracting a toxic or radioactive piece of material must be able to autonomously coordinate all the drones in the mission to identify and transport the material to the designated site. It is interesting when the decision-making process and the autonomy requirements are analysed against the framework proposed. Each drone in the fleet must be equipped with sensors that will enable it to identify the target material, fly in formation, and position appropriately to lift and extract the material. The multi-drone system in this scenario must be able to understand the mission and coordinate with other drones to fulfil it. However, this layer is the lowest in the hierarchy and can be overridden by layers 1 or 2.

6. Conclusions

The future of drones lies in the maturity and application of autonomous multi-drone systems capable of fulfilling different use cases. However, the definition of autonomy or autonomous capability is not settled for the multi-drone operational scenario. Whilst the push for the standardisation and approval for beyond visual line of sight (BVLOS) operations is important, it is equally necessary to have a common model or framework to guide BVLOS for multi-drone use in the civil aviation space. Multi-drone systems are very different from single drone uses although a significant aspect of current regulation is predicated or informed by single drone operational concepts. This chapter lays out a framework that may be of interest to regulators, UAV companies and the majority of other stakeholders. The framework defines autonomy in both single and multi-drone implementations in hierarchies and layers. The proposed multi-drone systems were segregated into 3 main layers and hierarchies of autonomy namely single vehicle control (Layer 1), multi-vehicle control (Layer 2) and global mission control (Layer 3). In this framework, Layer 1, which is single vehicle control, occupies the highest hierarchy and hosts functions performed by the flight control system (FCS) for instance. By developing and segregating these autonomy hierarchies, the design, implementation and regulation of autonomous multi-drone systems can be approached and managed efficiently across jurisdictions.

6.1 Future work

This is still a very dynamic area of research with multidisciplinary implications. As regulators request for input over the next few years, it is important to develop prototypes and demonstrations of the framework and ideas shared in this chapter. It is also important to look into how the physical forms of the drones may affect its operational environment/specific use cases and how that may impact autonomy considerations. Whilst the proposed framework is inherently designed to be agnostic to technology, it will be helpful to see how much technological developments will impact current thinking and by extension the extant regulatory regime.

Abbreviations

UA	Unmanned Aircraft
UAV	Unmanned Aerial Vehicle
UAS	Unmanned Aircraft System
VLOS	Visual Line of Sight
EVLOS	Extended Visual Line of Sight
BVLOS	Beyond Visual Line of Sight
PACT	Pilot Authority and Control of Tasks
ALFUS	Autonomous Levels For Unmanned Systems
EASA	European Union Aviation Safety Agency
FAA	Federal Aviation Authority
CAA	UK Civil Aviation Authority
ICAO	International Civil Aviation Organisation

Author details

Ogbonnaya Anicho
Liverpool Hope University, Liverpool, UK

*Address all correspondence to: anichoo@hope.ac.uk

References

[1] UKCAA. Decision Amending Acceptable Means of Compliance (AMC) and Guidance Material (GM) for UK Reg (eu) 2019/947 Regarding the Rules and Procedures for the Operation of Unmanned Aircraft. Technical report. London, United Kingdom: UK Civil Aviation Authority; 2022

[2] EASA. Easy Access Rules for Unmanned Aircraft Systems. Technical report. Cologne, Germany: European Union Aviation Safety Agency; 2022

[3] Sadraey M, H. Design of Unmanned Aerial Systems. New Jersey, United States: John Wiley & Sons Ltd; 2020

[4] Spouncer A. UAVs: Unlocking Positive Transformation in the World. Technical report. Cranfield, United Kingdom: Immarsat and Cranfield University; 2021

[5] Heutger M, Kückelhaus M. Unmanned Aerial Vehicles in Logistics: A DHL Perspective on Implications and Use Cases for the Logistics Industry. Technical report. Troisdorf, Germany: DHL Customer Solutions & Innovation; 2020

[6] Hatt T, Joiner J, Pareglio B. Race to the Top: Assessing and Accelerating Drone Readiness in the UK, the G7 and Other Leading Nations. Technical report. London, United Kingdom: BT Group and GSMA; 2023

[7] INSIGHT. The Journey Towards Autonomy in Civil Aerospace. Technical report. Cranfield, United Kingdom: Aerospace Technology Institute (ATI); 2020

[8] Chen H, Wang XM, Li Y. A Survey of Autonomous Control for UAV. Washington, D.C., United States: IEEE Computer Society; 2009

[9] Chen TB. Management of Multiple Heterogenous Unmanned Aerial Vehicles Through Capacity Transparency [thesis]. Queensland, Australia: Queensland University of Technology; 2016

[10] Ahmed H, Glasgow J. Swarm Intelligence: Concepts, Models and Applications. Technical report. Ontario, Canada: School of Computing, Queen's University Kingston; 2012

[11] NASA. NASAFacts: Mars Helicopter/Ingenuity. Technical report. California, United States: Jet Propulsion Laboratory - National Aeronautics and Space Administration (NASA); 2020

[12] UAV Navigation. UAS in Firefighting: How Unmanned Systems Can Help to Keep the Environment Safe. Technical report. Madrid, Spain: UAV Navigation; 2022

[13] Edward F. The Sustainability of Zipline's Autonomous Aerial Logistics. Technical report. California, United States: Zipline Drones; 2020

[14] Heath N. Project Wing: A Cheat Sheet on Alphabet's Drone Delivery Project. Technical report. Tennessee, United States: TechRepublic; 2018

[15] Elsight. Optimal Connectivity for Commercial BVLOS Drone Operations. White paper. Tel Aviv, Israel: Elsight Limited; 2023

[16] ICAO. ICAO Model UAS Regulations. Technical report. Montreal, Canada: International Civil Aviation Organisation (ICAO); 2020

[17] Elsight. BVLOS Across the Drone Industry. White paper. Tel Aviv, Israel: Elsight Limited; 2021

[18] FAA. 14 CFR Part 107 (up to date as of Aug. 9, 2023)Small Unmanned Aircraft Systems. Technical report. Washington, D.C., United States: US Federal Aviation Administration; 2023

Chapter 4

Deformable Correlation Networks for Aerial Object Tracking and Segmentation

Daitao Xing and Anthony Tzes

Abstract

While recent object trackers, which employ segmentation methods for bounding box estimation, have achieved significant advancements in tracking accuracy, they are still limited in their ability to accommodate geometric transformations. This limitation results in poor performance over long sequences in aerial object-tracking applications. To mitigate this problem, we propose a novel real-time tracking framework consisting of deformation modules. These modules model geometric variations and appearance changes at different levels for segmentation purposes. Specifically, the proposal deformation module produces a local tracking region by learning a geometric transformation from the previous state. By decomposing the target representation into templates corresponding to parts of the object, the kernel deformation module performs local cross-correlation in a computationally and parameter-efficient manner. Additionally, we introduce a mask deformation module to increase tracking flexibility by choosing the most important correlation kernels adaptively. Our final segmentation tracker achieves state-of-the-art performance on six tracking benchmarks, producing segmentation masks and rotated bounding boxes at over 60 frames per second.

Keywords: unmanned aerial vehicles, Siamese network, real-time object tracking, object segmentation, deep learning

1. Introduction

Visual object tracking (VOT) is a fundamental task in various applications, such as robot navigation [1], human-computer interaction [2], and unmanned aerial vehicle (UAV) based monitoring [3]. Given the initial state of an arbitrary target in the first frame, VOT aims to update the location and the states for all the subsequent video frames in real time. Nevertheless, the target object may undergo large appearance changes caused by illumination, deformation, occlusion, and fast motion, which make VOT a very challenging task.

Recent works [4, 5], inspired by the video object segmentation (VOS) task, attempt to obtain a precise target state from the binary per-pixel segmentation mask. A binary segmentation mask provides a much more detailed representation and a closer approximation of the target than a bounding box, especially for elongated and deformable objects. However, methods employing the u-net architecture or pixel-wise binary mask classification are still limited by the model's geometric transformations, resulting in unstable tracking performance. Among these methods, such as D3S and SATVOS, the focus is on reducing model complexity by utilizing prior knowledge to meet real-time requirements. Nevertheless, the segmentation is performed on the entire search region, which includes distractors. This approach hinders both robustness and speed.

In this research, we propose a novel real-time aerial object tracking framework that utilizes the accuracy of segmentation in a computationally and parameter-efficient manner. Compared to methods such as D3S [5] and SiamMask [4], which predict masks globally or densely, we perform segmentation only on a local tracking region. To achieve this, we introduce a proposal deformation module (PDM) to narrow down the search area to a fixed, small salience patch. Specifically, given the estimated object state from the last frame, the PDM module predicts geometric transformations of the target object by learning an offset from its old position, yielding a coarse updated target state. To better model target geometric variances, we propose a Kernel deformation module (KDM) that disentangles object pose and part deformation from texture and shape for segmentation. A target representation learned from the first frame is decomposed into templates corresponding to different parts of the object. These templates are then applied to the cropped patch from the PDM module in a pixel-wise manner. With a reduced segmentation area, fewer background distractors, and reduced model complexity, our proposed modules achieve higher segmentation accuracy and speed.

To further improve the model's capability of adaption to the appearance variance and abnormal states like occlusion and truncation, we employ the binary mask from the initial frame and propose a mask deformation module (MDM).

Inspired by Spatial Transformer Networks (STN) [6], an affine transformation function is regressed with a fully connected network conditioned on the current search frame without additional supervision. The transformation is then performed on the initial mask and can include scaling, cropping, and rotations as well as non-rigid deformations. This endows the network with the flexibility of choosing the most important correlation kernels adaptively, resulting in an improvement in model generalization and robustness.

We conducted comprehensive experiments on multiple benchmarks, including six challenging VOT datasets: VOT2016 [7], VOT2018 [8], OTB100 [9], TrackingNet [10], Got10K [11], and LaSOT [12], as well as two VOS datasets: DAVIS16 [13] and DAVIS17 [14]. Superior performance, as well as extensive ablation studies, demonstrate the effectiveness of the proposed method. Particularly, our approach achieves state-of-the-art performance on VOT benchmarks, with an EAO score of 0.485 on VOT2018 [8], while running at over 60 FPS. While our tracker was not originally designed for the VOS task, it still achieves compatible results on both VOS benchmarks.

2. Related work

2.1 Visual object tracking

SiamFC introduces a Siamese architecture, which measures the similarity between the target and search image and trains the network offline. SiamRPN [15] introduces the region proposal network to jointly perform classification and regression in an end-to-end way, DaSiamRPN [16] improves the discrimination power of the model with a distractor-aware module, and SiamRPN++ [17] further improves performance with more powerful deep architectures by overcoming spatial invariance restrictions. Recent works like SiamFC++ [18], Ocean [19], and SiamBAN [20] replace the RPN with an anchor-free mechanism and achieve faster tracking speed. Siamese network-based trackers rely on the first frame to learn a robust target appearance representation without adaptation, which is unreliable during long-term tracking. In contrast to Siamese-based approaches, recent works like DiMP [21] and ATOM [22] learn a discriminative classifier online to distinguish the target from the background. However, the adaptation exists only for classification, while the bounding box estimation still relies on a fixed template matching strategy.

2.2 Video object segmentation

Most recent works [23–26] resort to large deep networks, multi-scale feature fusion, u-net architecture, or fine-tuning techniques, achieving impressive accuracy. However, tracking efficiency is often overlooked, which is crucial for generic object tracking. Inspired by the significant advancements in the visual object tracking task, recent works tend to solve VOS with a correlation mechanism. SiamMask [4] first proposed a unified approach that predicts the bounding box and mask from a correlation feature map simultaneously. SiamRCNN deploys the segmentation by detection paradigm and predicts the mask inside each proposed box. SiamAtt [27] improves the robustness and discriminative representation ability of the correlation feature map by introducing an attention mechanism. It then performs state estimation and segmentation independently on a refined proposed region. D3S [5] removes the bounding box regression branch and performs segmentation based on the classification map and similarity score map, achieving state-of-the-art performance on the VOT2018 [8] benchmark. However, all these works rely on multiple refinement modules made of upsampling layers and skipping connections to increase robustness, which can also hinder tracking efficiency.

2.3 Transformation modeling

Spatial transformer networks (STN) [6] were the first work to improve the translation-invariant feature modeling ability for deep CNNs by learning a transformation affine function from the input image. Instead of performing global affine transformations, Deformable ConvNets [28] learn offsets for each convolution kernel or ROI pooling layer in a dense way, achieving significant performance improvement on complex vision tasks.

3. Method

In this section, we describe our tracking framework in detail. The proposed method DCNet consists of four parts: a Siamese-based backbone network for feature extraction presented in Section 3.1, a proposal deformation module (PDM) presented in Section 3.2, a Kernel deformation module (KDM) presented in Section 3.3, and a mask deformation module (MDM) presented in Section 3.4. **Figure 1** describes the architecture outline.

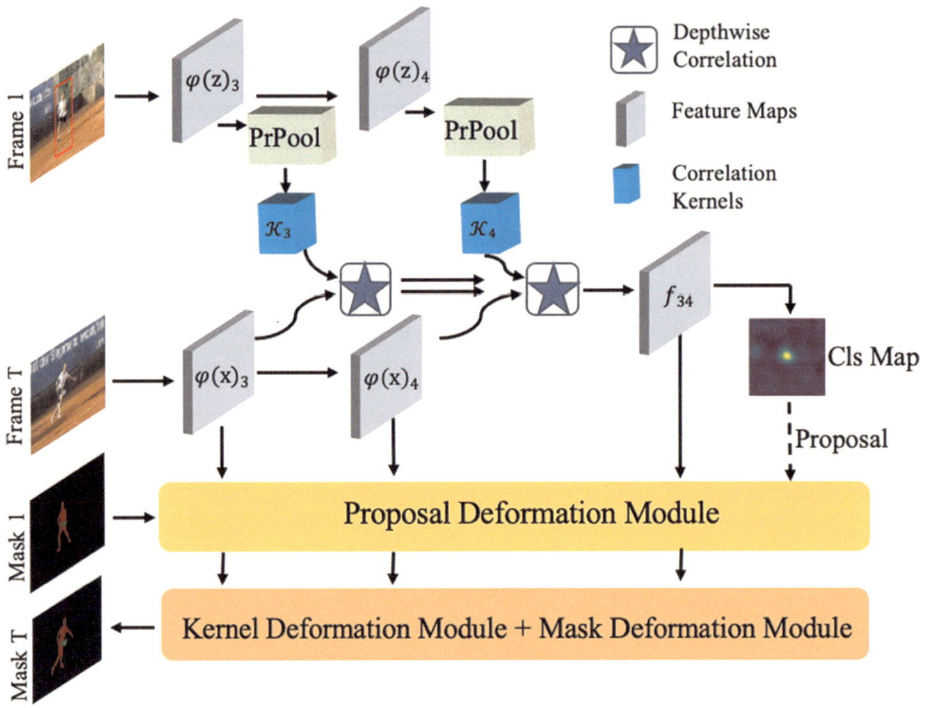

Figure 1.
An overview of the proposed tracking framework, consisting of an Siamese-based correlation network, and three deformation modules. The correlation kernels are cropped from different feature blocks, following PrPooling layer.

3.1 Network architecture

Before detailing the proposed modules, we first briefly introduce the backbone network. Similar to a Siamese tracking framework, the proposed framework DCNet consists of two branches: the template branch which takes cropped image z of size $w \times h$ from the initial frame as a reference, and the search branch which takes the cropped image x of the same size with z from the current frame for tracking. The two inputs are processed by the same backbone network ψ, yielding two feature maps $\psi(z) \subset \mathbb{R}^{W \times H \times C}$ and $\psi(x) \in \mathbb{R}^{W \times H \times C}$ for matching purposes. The reference image also provides the initial target state, represented as a bounding box $\mathcal{B}_0 = (c_0^x, c_0^y, w_0, h_0)$. An alternative representation, a binary mask $\mathcal{M}_0 \in \mathbb{R}^{w \times h \times 1}$ is given in some scenarios but is nonessential.

As in most Siamese-based trackers, cross-correlation is performed between feature maps $\psi(z)$ and $\psi(x)$. The response score maps representing the similarity between the reference image and the test image are used for target localization and state estimation. We thus retrieve two object-aware correlation kernels $\mathcal{K} \in \mathbb{R}^{6 \times 6 \times C}$ from $\psi(z)$ to represent target appearance. Specifically, we crop patches from *block3* $\psi(z)_3$ and *block4* $\psi(z)_4$ with the initial state $\mathcal{B}_0$ and feed them into a Precise ROI Pooling (PrPool) [29] layer, yielding two fixed-size feature maps $\mathcal{K}_3$ and $\mathcal{K}_4$. Finally, a depthwise correlation is performed between $\mathcal{K}$ and $\psi(x)$ to generate a target-aware feature representation $f_{3,4} \in \mathbb{R}^{W \times H \times C}$. Following SiamRPN++ [17], we deploy the multi-layer aggregation strategy by concatenating features $f_{3,4}$ and feed the joined feature map f_{34} into a classification module for target localization.

3.2 Proposal deformation module

Compared to D3S [5] and SiamMask [4] which predict masks globally or densely, we only perform prediction on a local tracking region. However, accurate and fast segmentation of a target with appearance changes and deformation requires a proper segment input that encloses the entire target object while precluding the distractors from the background. While the target-aware feature f after correlation can be used to distinguish the target from background distractors, it does not adapt to appearance changes, leading to reduced robustness in localization. To this end, we first propose the proposal deformation module (PDM) to reduce the size of segment input to the ideal area by deforming the region of interest (RoI) from the search area. Unlike the region proposal network (RPN) from mask-RCNN, we make use of the inter-frame consistency of video and generate RoI from the target state in the last frame. Specifically, given the maximum of the correlation response, i.e., $\hat{c}_t^x, \hat{c}_t^y$, together with the width and height of the estimated state from the last frame, we generate an initial RoI represented as $\mathcal{R} = \left(\hat{c}_t^x, \hat{c}_t^y, \hat{w}_{t-1}, \hat{h}_{t-1}\right)$.

Proposal deformation: As shown in **Figure 2**, the proposed module refines by $\mathcal{R}$ learning a deformation from the initial RoI to the ground-truth state. A PrPool layer is applied on target-aware feature f_{34}, yielding a pooled representation of size $K \times K \times C$, where K is the spatial output size of the PrPool layer. From the feature representation, a fully connected layer generates the offsets d, which are then added to the initial RoI $\mathcal{R}$, resulting in a refined RoI $\mathcal{R}'$. We represent the offsets $\boldsymbol{d} = \left(lt^h, lt^w, rb^h, rb^w\right)$ as the distances from $\mathcal{R}$'s left-top corner and right-bottom corner to the corresponding ground-truth ones. Thus, d can be formulated as:

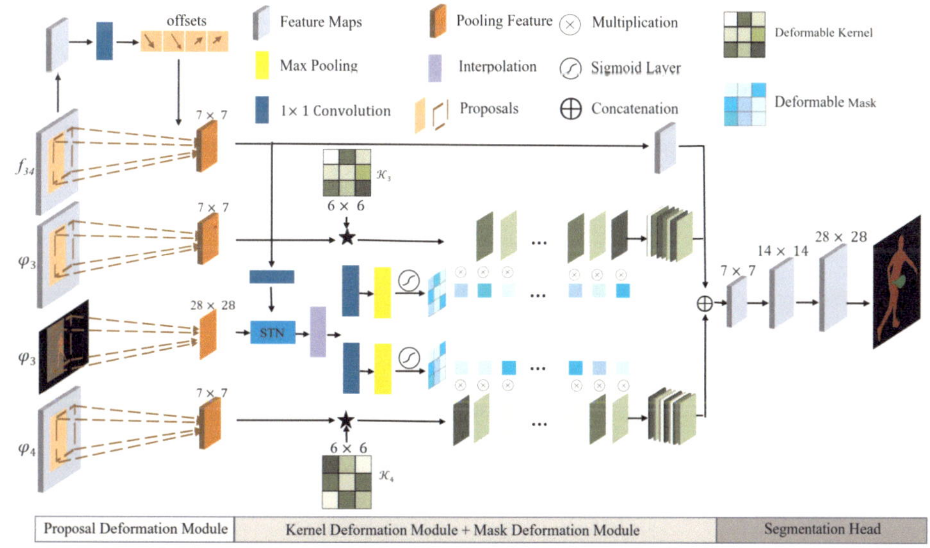

Figure 2.
The proposed deformation modules. The PDM takes the fusion correlation feature and initial proposal as inputs, predicts the offset, and yields a new proposal. The KDM performs part similarity learning (⋆) between the ROI feature and kernel sets from the initial frame. Further, a deformed mask learned from spatial transformation network assigns different scores to the corresponding correlation map adaptively. The final segmentation is inferred from the feature map consisting of local and global correlation maps.

$$\begin{aligned} lt^h &= \left(c_t^y - \frac{h_t}{2}\right) - \left(\hat{c}_t^y - \frac{h_{t-1}}{2}\right) \\ lt^w &= \left(c_t^x - \frac{w_t}{2}\right) - \left(\hat{c}_t^x - \frac{w_{t-1}}{2}\right) \\ rb^h &= \left(c_t^y + \frac{h_t}{2}\right) - \left(\hat{c}_t^y + \frac{h_{t-1}}{2}\right) \\ rb^w &= \left(c_t^x + \frac{w_t}{2}\right) - \left(\hat{c}_t^x + \frac{w_{t-1}}{2}\right) \end{aligned} \tag{1}$$

where c_t^y, c_t^x, h_t and w_t are from the ground-truth state $\mathcal{B}_t$. Since $\mathcal{R}$ provides a good approximation of the target state, the module is lightweight with only a small amount of parameters and computations for the offset learning.

3.3 Kernel deformation module

Given the refined RoI $\mathcal{R}'$, a small feature map $\mathcal{H}$ of size 7×7 is extracted with a PrPool layer from feature maps $\psi(x)_{3,4}$. Generally, the pooled feature map can be used for instance segmentation in object detection scenarios like MaskRCNN [30]. For the generic object tracking task, however, the object class is nonessential to belong to any set of known categories. Thus, the segmentation introduces misalignment between the target and the instance object, which has a large negative effect on the prediction of masks with pixel accuracy. While cross-correlation can be used to determine interesting parts for segmentation, the cross-correlation methods in [30] lack the ability to model the geometric variance of the target. To this end, we propose a Kernel Deformation Module (KDM) which utilizes part correlation for segmentation purposes.

3.3.1 Part correlation

Formally, given the compact target representation $\mathcal{K}$ of size 6×6 from $\psi(z)_{3,4}$ which contains local information (e.g. color, shape) of the target, we first decompose the kernel $\mathcal{K}$ into a separate vector kernel set $\{\mathcal{K}_i | i \in \{0, \ldots 36\}\}$. The cross-correlation is performed between $\mathcal{H}$ and kernel set $\mathcal{K}_i$ which can be formulated as:

$$\mathcal{C} = \{\mathcal{C}_i | \mathcal{C}_i = \mathcal{K}_i \star \mathcal{H}\}_{i \in \{0, \ldots 36\}} \tag{2}$$

where $\mathcal{C}_i \in \mathbb{R}^{7\times7\times1}$ is the response attention map after correlation with kernel $\mathcal{K}_i$. Apparently, kernels that represent distinct local information activate disparate regions on the feature maps. This allows similarity learning which includes object part correlation to find regions of a search image that are most relevant (attention). Finally, we concatenate attention maps set $\mathcal{C}$ into a union feature map $\mathcal{C}' \in \mathbb{R}^{7\times7\times36}$.

3.3.2 Segmentation head

With the aim of producing a more accurate object mask, we merge low and high-resolution features using parallel part-matching branches. As shown in **Figure 2**, we use both Block3 and Block4 from $\psi(x)$ and generate two pooled feature maps $\mathcal{H}_3, \mathcal{H}_4$, resulting in two correspondent maps $\mathcal{C}_3', \mathcal{C}_4'$. These two attention maps are concatenated before being fed into the segmentation branch. To further improve the

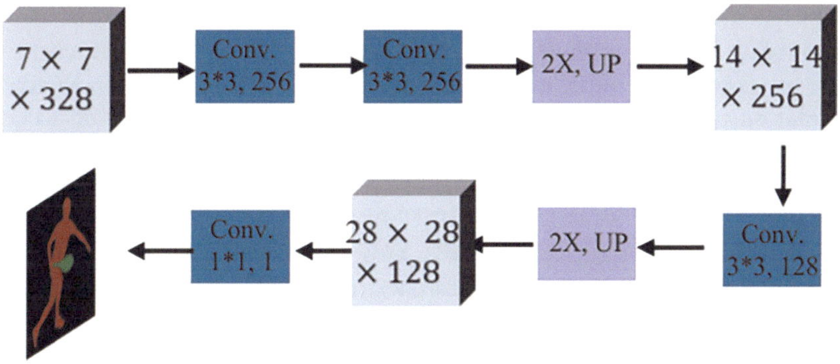

Figure 3.
Architecture details of segmentation head.

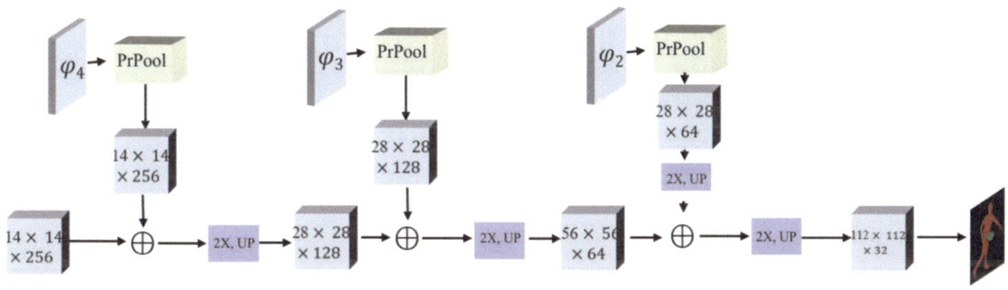

Figure 4.
Architecture details of refine module.

discriminative representation ability, we also crop a feature map of size $7 \times 7 f_{34}$ from and attach it with the attention maps. For the segmentation head, we closely follow the architectures presented in previous work [30]. Specifically, three upsampling blocks consisting of one convolution layer, one activation function, and one deconvolution layer are connected and map the input $(\mathcal{C}'_3 \| \mathcal{C}'_4)$ into a predicted mask $\mathcal{M}_t \in \mathbb{R}^{28\times28\times1}$. **Figure 3** shows the architecture of the segmentation head. We follow the design from MaskRCNN [30]. Specifically, we use two interpolation layers with convolution layers to up-sample the target into $56 * 56$ for end-to-end training. During inference, the output mask is resized to the original size. For video segmentation tasks, the pixel-wise segmentation performance is highly related to the size of the segmentation mask. We have since designed a refined module, following the skip connection strategies [4]. As shown in **Figure 4**, we merge low and high-resolution features by cropping and pooling different blocks (ψ_2 to ψ_4), and enlarging the size of mask so that the output includes enough details.

3.4 Mask deformation module

While the KDM is able to accommodate geometric transformations in object pose, scale, and part deformation, the fixed correlation template strategy without adaptation, hinders the tracking robustness for target appearance variance in the long term, especially in abnormal states like occlusion and truncation. To further improve tracking robustness in harsh cases, we utilize the binary mask from the initial frame $\mathcal{M}_0$

and propose a mask deformation module. Specifically, we equip the initial sampling grid from $\mathcal{M}_0 \in \mathbb{R}^{28\times28\times1}$ with a spatial transformation network $\mathcal{T}$ to convert the sampling mask from the initial state to the deformed state. As shown in **Figure 2**, the transformation function is $\mathcal{T}$ obtained from the target-aware map f using a PrPool followed by a fully connected layer. The transformed mask is further encoded with two 3 × 3 convolutional layers, following a max pooling on feature depth to reduce channels into one. The feature map is turned into the desired shape by interpolation, yielding the final encoded mask $\hat{\mathcal{M}}$. While $\hat{\mathcal{M}}$ is a coarse map with limited details provided, the value on each sampling position indicates the importance of the corresponding part from the correlation kernel $\mathcal{K}$. Thus, the part matching could be updated as:

$$\mathcal{C} = \{\mathcal{C}_i | \mathcal{C}_i = \sigma(m_i) * \mathcal{K}_i \star \mathcal{H}\}_{i\in\{0,\dots 36\}} \tag{3}$$

where m_i is the scalar value from the transformed mask $\hat{\mathcal{M}}$ and σ denotes the sigmoid function which turn m_i into a significant score. This endows the network with the flexibility of choosing the most important correlation kernels adaptively. Since both the grid sampling and transformation are differentiable, no supervision is required to learn the transformation parameters θ. In some scenarios such as motion blur or "non-object" instance where a segmentation mask is not provided or unavailable, the mask $\mathcal{M}_0$ is filled with ones that gives equal importance to part-matching kernels.

4. Discriminative segmentation tracker

This section describes the implementation details from training to online tracking with the proposed DCNet method.

4.1 Training

Similar to the Siamese-based tracker, the proposed network is trained offline with image pairs. For considerable tracking robustness and speed, we use an ImageNet [31] pre-trained ResNet-50 [32] model as a backbone network. We use the same training split as ATOM [22], i.e., LaSOT [12], TrackingNet [10], and COCO [33] datasets. We also include YouTube-VOS [34] for segmentation training purposes. Like SiamRPN [15], we sample image pairs from the videos with a maximum gap of 50 frames and augment the image dataset (e.g., COCO) with random transformation to generate synthetic image pairs. From a reference image, we sample a square patch centered at the target, with an area of 5^2 times the target size. For the test image, we sample a similar patch, with some perturbation in the position and scale. Both patches are resized to a fixed size of 288 × 288. To simulate the tracking scenario, we first generate a region proposal by adding random Gaussian noises to the ground truth bounding box and generating 16 candidate states. The variance of the Gaussian function is sampled randomly from [0.01, 0.05, 0.1, 0.2, 0.3, 0.4].

We set $C = 256$ for feature representation in transformation modules. For the PDM, the spatial output size K of the PrPool layer is set to 7. The fully connected layer in MDM used to learn the transformation parameters has a weight matrix filled with zeros and a bias equal to [1, 0, 0, 0, 1, 0].

4.2 Mask representation

In contrast to segmentation methods [5] in the style of FCN [35], which predicts a mask on the entire search area, our approach restricts the segmentation inside the RoI area for inference speed improvement. Apparently, there is an estimation bias of the RoI proposal before and after refinement in PDM, which may impede the training process, resulting in an upper bound of the segmentation head. To address this problem, we expand the proposal size by manipulating a factor γ to the width and height of RoI. To prevent overfitting when learning the segmentation scale changes, we uniformly sample a factor between 1.2 and 1.5 during training. For the inference stage , is fixed to 1.4 for all scenarios. we argue that this process will not introduce additional computation since the RoI is pooled into fixed size, but stabilizes the training process and improves the performance.

Unlike the VOS task, the initial mask is not provided in the VOT dataset. Besides, some scenarios cannot provide a valid semantic mask, which disturbs the segmentation process during inference. We thus fill the initial mask with ones given a probability for the training stage. For inference, an estimated binary mask is predicted using GrabCut [36] in case the mask is not provided. If the estimation fails for abnormal cases, a mask of the same size with the bounding box, filled with ones will be used in MDM.

4.3 Training objective

Both PDM and KDM require supervision during training. We use the Smooth L1 loss for offset learning in Eq. 1 and Binary Cross Entropy (BCE) loss for both binary mask generation and target localization. The total loss of DCNet is:

$$L = \lambda_1 L_{cls} \pm \lambda_2 L_{reg} + \lambda_3 L_{mask} \tag{4}$$

where $\lambda_{1,2,3}$ are weights for balancing losses and $\lambda_1 = \lambda_2 = \lambda_3 = 1$. Since the initial RoI proposals have a relatively high IoU score between each other, it is unnecessary to predict masks for each RoI proposal. During training, a random proposal is chosen to train the segmentation module. For inference, instead, we use the default proposal without augmentation. The network is trained for 40 epoches with 32 image pairs per batch, giving a total training time of less than 1 day on four GPU TITAN X servers. We use ADAM [37] optimizer with an initial learning rate of 10^{-3}, and decay of the learning rate with a factor of 0.9 every epoch.

4.4 Box generation

We consider the same strategy as [4] to generate a bounding box from a mask. Specifically, a mask from the segmentation branch is converted to a binary mask with a threshold of 0.5. Then, the rotated minimum bounding rectangle (MBR) is applied to yield a rotated bounding box. The segmentation may fail in some abnormal states where the target loses the semantics and cannot be discriminated from the background. We thus measure the IoU score between the bounding box estimation from the segmentation branch with the refined RoI proposal and use the RoI value as the final prediction if the IoU score is less than 0.5. Thanks to the efficiency and robustness of the proposed modules, no additional post-processing steps are required to get the final prediction, yielding a fast inference speed of over 60 FPS on an Nvidia 2060 GPU.

5. Experiments

In this section, we compare our approach with the state-of-the-art (sota) methods in two related tasks: VOS and VOT. We first evaluate DCNet on six challenging tracking benchmarks: VOT2016 [7], VOT2018 [8], OTB100 [9], TrackingNet [10], Got10K [11], and LaSOT [12] in Section 5.1. To extensively evaluate the proposed method, we compare it with 21 state-of-the-art trackers. There are 4 anchor-based Siamese framework based methods (SiamRPN [15], SiamRPN++ [17], DaSiamRPN [16] and UpdateNet [38]),4 anchor-free Siamese framework based methods (SiamFC [39], SiamFC++ [18], SiamBAN [20] and Ocean [19]),7 discriminative correlation filter based methods (ECO [40], CSRDCF [41], CCOT [42], ASRCF [43], ROAM [44], DiMP [21], and ATOM [22]), 2 attention based methods (SiamAttn [27], CGACD [45]), 1 multi-domain method (MDNet [46]) and 3 tracking by segmentation methods (D3S [5], SiamMask [4] and SiamRCNN [47]). We then perform an analysis of the segmentation capabilities of DCNet on two popular VOS benchmarks: DAVIS2016 [13] and DAVIS2017 [14] in Section 5.2. Finally, an ablation study is performed based on the VOT2018 [8] dataset in Section 5.3.

5.1 Evaluation on VOT benchmarks

VOT2016 [7] and VOT2018 [8] VOT datasets each consist of 60 sequences with different challenging factors in which targets are annotated by rotated rectangles. VOT benchmarks use the protocol [7] in which the tracker is reset upon tracking failure. Therefore, the tracking results are more comprehensive and insightful. The performance is compared in terms of accuracy (average overlap over successfully tracked frames), robustness (failure rate), and the EAO (expected average overlap), which is a principled combination of the former two measures.

In the VOT2016 dataset, we compare DCNet with the following sota trackers which cover most of the current representative methods: two segmentation-based trackers D3S [5], SiamMask [4], Siamese network-based trackers DaSiamRPN [16], SiamRPN [15], discriminative correlation filter based methods ROAM [44], ASRCF [43], and CSRDCF [41].

Results are reported in **Table 1**, DCNet outperforms all top-performance sota by a large margin especially in EAO measurement, achieving approximately a 6% boost in EAO compared to the D3S and almost 20% boost in EAO compared to the SiamMask. The accuracy inference comes from two aspects: (1) the segmentation is performed on the RoI proposal instead of the entire search area which may cause truncation if the RoI is not precise enough. (2) The output of segmentation is limited to 28 $\times$ 28 for best speed and accuracy trade-off. Nevertheless, the efficiency of the proposed transformation module brings both robustness and speed to the tracker, resulting in an overall

	CSRDCF [41]	DaSiam RPN [16]	Siam RPN [15]	ASRCF [43]	Siam Mask [4]	ROAM [44]	D3S [5]	Ours
A	0.51	0.610	0.560	0.563	0.670	0.599	0.660	0.640
R	0.238	0.220	0.302	0.187	0.233	0.174	0.131	0.125
EAO	0.338	0.411	0.344	0.391	0.442	0.441	0.493	0.526

Table 1.
VOT2016 [7] – Comparison with state-of-the-art trackers.

	Siam RPN++ [17]	DiMP [21]	ATOM [22]	Siam FC++ [18]	Siam Mask [4]	Siam Attn [27]	Siam RCNN [47]	D3S [5]	Ours
A	0.600	0.597	0.590	0.587	0.642	0.630	0.609	0.640	0.602
R	0.234	0.153	0.204	0.183	0.295	0.16	0.220	0.150	0.145
EAO	0.414	0.440	0.401	0.426	0.387	0.470	0.408	0.489	0.485

Table 2.
VOT2018 [8] – Comparison with state-of-the-art trackers.

boost in performance. Compared with D3S which runs only at 25 FPS, our approach doubles the inference speed, achieving 62 FPS with a significant accuracy increase.

Table 2 shows the performance results on the VOT2018 [8] dataset. DCNet still achieves sota performance in all measures, compared with the recent top-performer D3S. Our method achieves an EAO score of 0.485, a relative 25% improvement compared with SiamMask. Furthermore, DCNet achieves a top robustness score of 0.145 and outperforms the most recent sota discriminative correlation filter DiMP [21] and ATOM [22], demonstrating the robustness of our state transformation module.

OTB100 [9] The OTB100 [9] dataset contains 100 fully annotated video sequences with substantial variations. We adopt the straightforward One-Pass Evaluation (OPE) as the performance evaluation method. For the performance evaluation metrics, we use precision plots and success plots. Following the protocol in the OTB benchmark, we use the threshold of 20 pixels and Area Under Curve (AUC) to present and compare the representative precision plots and success plots of trackers, respectively. **Figure 5** shows the success and precision plots on OTB100. DCNet achieves a similar AUC score compared with top performer SiamRPN++ [17] and outperforms all other sota methods SiamPRN++ [17]. Specifically, our DCNet surpasses ATOM, which is the top discriminative tracker using the same online classification module for target localization, by 4.5 and 4.8% on the success and precision plots respectively, demonstrating the effectiveness of refining RoI proposals using the state transformation module. In addition, our DCNet outperforms the very recent anchor-free tracker SiamBAN [20] on the precision plot.

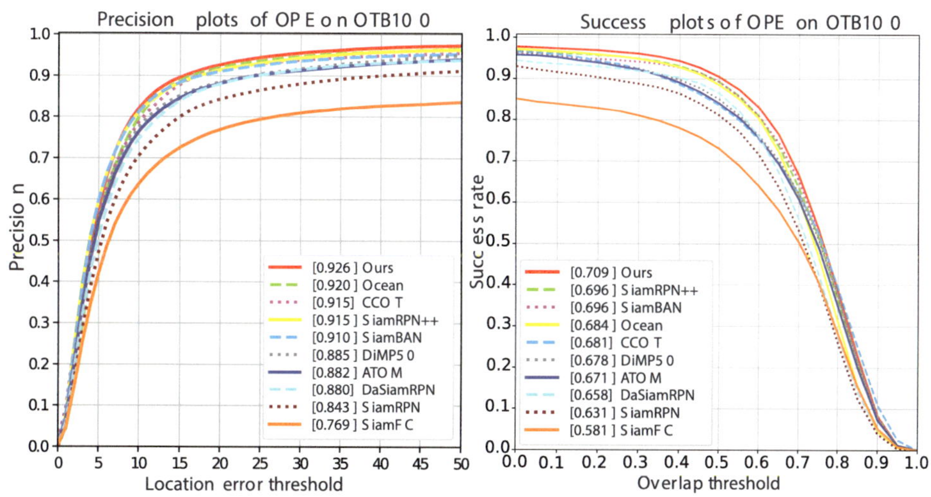

Figure 5.
Evaluation results of trackers on OTB100 [9].

	Update Net [38]	DiMP [21]	ATOM [22]	Siam RPN++ [17]	CGACD [45]	ROAM [44]	PrDiMP [48]	Ours
Succ.	0.677	0.740	0.703	0.733	0.711	0.670	0.758	0.729
Prec.	0.625	0.687	0.648	0.694	0.693	0.623	0.704	0.693

Table 3.
Results on the TrackingNet [10] test set in terms of precision and success (AUC).

TrackingNet [10] TrackingNet [10] is a large-scale dataset consisting of more than 30 K videos for training and 511 videos for testing. Trackers are ranked in terms of AUC, and precision. As shown in **Table 3**, DCNet significantly outperforms all sota methods on precision metric. In detail, our DCNet obtains an improvement of 4.4, 7.3, and 1.1% on precision score compared with top-performing trackers D3S, ATOM [22], and DiMP [21], respectively.

Got10K [11] Got10K [11] is another large-scale dataset that was proposed recently. Following the requirement of generic object tracking, there is no overlap in object categories between the training set and test set, which is more challenging and requires a tracker with a powerful generalization ability. We follow their protocol and train the network with training split and YouTube-VOS [34] dataset. The results are shown in **Table 4**. Note that all the results are tested on their online server. Trackers are ranked according to the average overlap and success rates at two overlap thresholds 0.5 ($SR_{0.5}$) and ($SR_{0.75}$) 0.75, respectively. We consider the following sota trackers in comparison: D3S [5], ROAM [44], DiMP [21], SiamMask [4], SiamFC++ [18], ATOM [22], and MDNet [46]. DCNet outperforms all sota methods by a large margin in all performance measures, achieving a relative 4% higher performance in average overlap compared with the sota Siamese-based tracker SiamFC++ [18]. It also outperforms the most recent ATOM [22] and D3S [5] by over 3.7% and over 10.9% in average overlap, respectively, demonstrating the powerful capability of generalized object tracking.

LaSOT [12] LaSOT [12] contains 280 video sequences for testing and employs success and precision plots as measurements. We evaluate our DCNet against the top-performing trackers, including Ocean [19], ATOM [22], DiMP [21], SiamRPN++ [17], SiamFC [39], ECO [40], and SiamBAN [20]. **Figure 6** presents the comparison results. Our DCNet achieves sota performance compared with the recent DiMP [21]. Specifically, our DCNet achieves an AUC score of 0.570, with a boost of over 52.8% on performance since MDNet.

NFS30NFS dataset contains 100 fully annotated video sequences with substantial variations. For the performance evaluation metrics, we use precision plots and success plots. The performance of DCNet is compared with sota trackers including SiamBAN [20], ATOM [22], DiMP [21], UPDT [49], ECO [40], MDNet [46], and CCOT [42]. **Figure 7** shows the success and precision plots on NFS30. DCNet outperforms all

	DiMP [21]	Siam FC++ [18]	ROAM [44]	D3S [5]	Ocean [19]	Ours
AO	0.611	0.596	0.465	0.597	0.611	0.618
SR0.5	0.717	0.694	0.532	0.676	0.721	0.734
SR0.7	0.492	0.479	0.236	0.462	—	0.488

Table 4.
State-of-the-art comparison on the GOT-10 k test set in terms of average overlap (AO) and success rates (SR) at overlap thresholds 0.5 and 0.75.

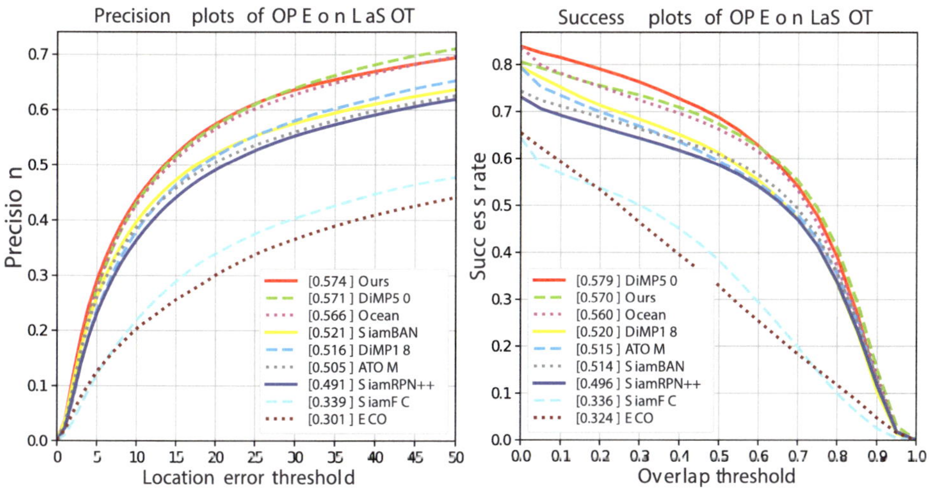

Figure 6.
Evaluation results of trackers on LaSOT [12].

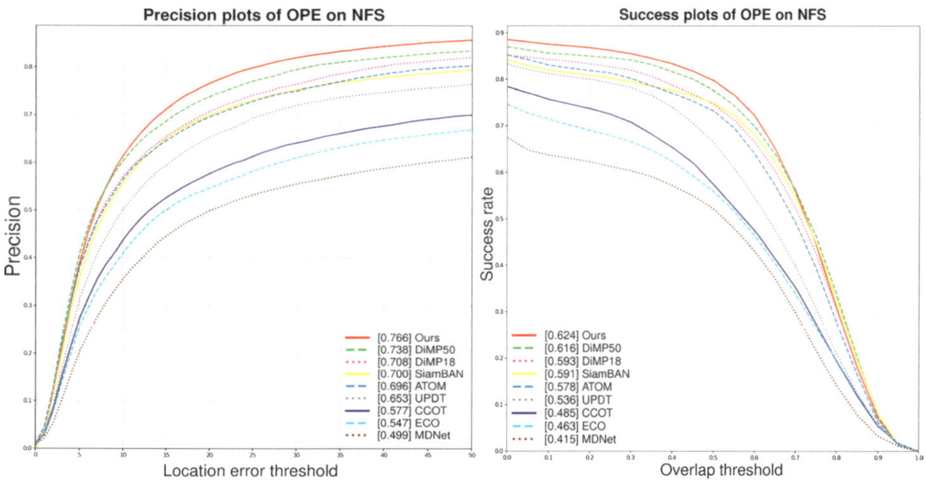

Figure 7.
State-of-the-art comparison on the NFS dataset in terms of precision and success rate.

state-of-the-art trackers under both metrics. Specifically, our DCNet surpasses DIMP-50, which is the top discriminative tracker using the online classification module for target localization, by 4.1% on the precision plots, while achieving a better success rate score, demonstrating the effectiveness of refining RoI proposals using the state deformation module. In addition, our DCNet outperforms the very recent anchor-free tracker SiamBAN [20] on both precision and success plots.

5.2 Evaluation on VOS benchmarks

For both DAVIS datasets, we use the official performance measures: the Jaccard index ($\mathcal{J}$) to indicate region similarity and the F-measure ($\mathcal{F}$) to represent contour

	DAVIS2016		DAVIS2017		FPS
	$\mathcal{J}_\mathcal{M}$	$\mathcal{F}_\mathcal{M}$	$\mathcal{J}_\mathcal{M}$	$\mathcal{F}_\mathcal{M}$	
D3S [5]	75.4	72.6	57.8	63.8	25.0
SiamMask [4]	71.7	67.8	54.3	58.5	55.0
OnAVOS [23]	86.1	84.9	61.6	69.1	0.1
FAVOS [24]	82.4	79.5	54.6	61.8	0.8
OSVOS [25]	79.8	80.6	56.6	63.9	0.1
DCNet-w/o mask	72.9	65.7	53.2	56.1	62.0
DCNet	74.5	71.1	54.2	59.4	62.0
DCNet-refine	76.9	73.2	57.9	63.3	28.0

Table 5.
State-of-the-art comparison over the DAVIS2016 and DAVIS2017 VOS benchmarks. Mean Jaccard index and mean F-measure are donated as $\mathcal{J}_\mathcal{M}$ and $\mathcal{F}_\mathcal{M}$, respectively.

accuracy. Following D3S [5], we only report the mean Jaccard index ($\mathcal{J}_\mathcal{M}$) and mean F-measure. We compare DCNet with several sota VOS methods including D3S [5], SiamMask [4], OnAVOS [23], FAVOS [24], and OSVOS [25].

For a compact comparison, we designed three variants, i.e. (1) DCNet without mask as input in which case $\mathcal{M}_0$ the is filled with (2) The proposed method with both bounding box and mask as input. (3) The proposed method follows with refinement modules as in [5]. **Table 5** shows the results of the comparison. DCNet achieves a compatible performance compared with segmentation-based trackers D3S [5] and SiamMask [4]. Compared to D3S [5], the performance of DCNet-w/o mask on DAVIS2016 [13] is only 3 and 9% lower in the mean Jaccard index and mean F-measure respectively, but with speed improvement over 1.5 times. A relatively smaller difference is achieved when providing the tracker with an initial mask. Due to the low resolution (28×28) of the segmentation mask, the KDM is prone to giving a lower performance score, especially for the F-measure metric, which expresses the contours similarity. By attaching a refinement module, DCNet outperforms the performance of D3S and SiamMask on DAVIS2016 [13] while reducing the inference speed by over 55%. We thus do not choose to extend the size of the segmentation output. This choice contributes significantly to the efficiency which is crucial for the visual tracking task.

5.3 Ablation analysis

An ablation analysis is performed on the VOT2018 [8] dataset to prove the effectiveness of the different modules of DCNet framework. We establish a baseline by removing all three transformation modules. Specifically, we eliminate the PDM and MDM modules and perform segmentation on the expanded RoI region with the scale factor 0.5 using FCN [35] style, following D3S [5]. The output of the segmentation head has full resolution and same size as the input. The results of the ablation study are presented in **Table 6**.

Effectiveness of PDM. Adding PDM to the baseline model causes an 11.5% performance gain with an ignorable impact on inference speed. This demonstrates the PDM has the capability of locating the most relevant regions of an image in a computationally and parameter-efficient way. **Figure 8** shows the state transformation process with RoI proposals before and after refinement.

	PDM	KDM	MDM	EAO	FPS
Baseline	—	—	—	0.397	40
+ PDM	✓	—	—	0.442	38
+ KDM	✓	✓	—	0.477	65
+ MDM	✓	✓	✓	0.484	62

Table 6.
Ablation analysis of DCNet on VOT2018 [8] dataset. We compare the impact of proposal deformation module (PDM), kernel deformation module (KDM), and mask deformation module (MDM).

Effectiveness of KDM. The overall tracking performance increased by 7.7% with KDM included to replace the FCN architecture. We argue that the improvement is mainly attributed to two aspects: (1) The FCN [35] relies on skipping connection mechanisms to replenish details, but introduces additional parameters to optimize, causing overfitting problem. We only perform segmentation on a low-resolution patch making the network much easier to converge. (2) FCN deploys a template matching strategy which makes the prediction unstable in abnormal cases over long sequences. Instead, KDM is able to accommodate model geometric transformation by using a part-matching strategy. The attention layers from different part matching are presented in **Figure 9**.

Figure 8.
Impact of PDM, green and red rectangles are RoIs before and after refinement, respectively.

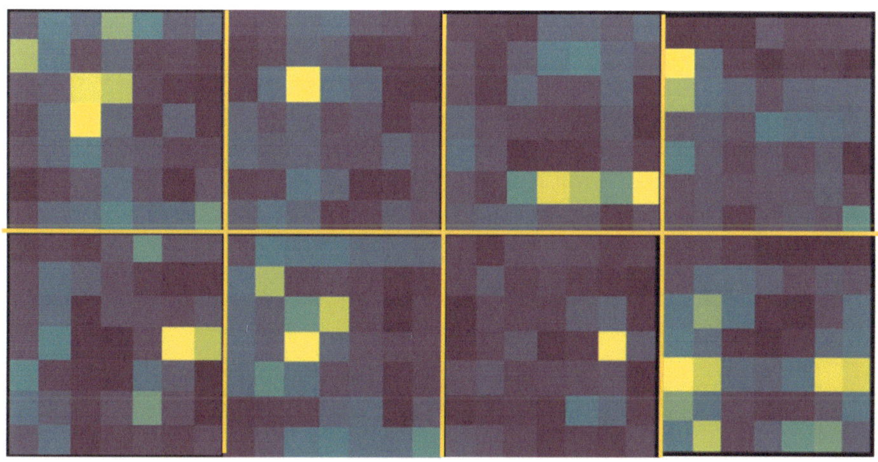

Figure 9.
Impact of KDM, attention maps from different part matching results.

Figure 10.
Segmentation mask and final output: Predicted mask from segmentation branch after interpolation and final output with rotated bounding and mask.

Effectiveness of MDM. MDM assigns different weights to part kernels in KDM which is helpful in some abnormal cases like occlusion and truncation. While only a small fraction of those extreme cases appear over the whole video sequence, they become pivotal when designing a robust tracking framework, as shown in **Figure 10**. Overall, MDM brings an extra 1.4% improvement in EAO score with a slight speed drop.

Speed Analysis In **Figure 11**, we visualize the EAO on VOT2018 with respect to the Frames-Per-Second (FPS). From the plot, our method achieves the best performance, while still running at real-time speed (62 FPS). **Figure 10** shows further qualitative results of segmentation branch and bounding box estimation.

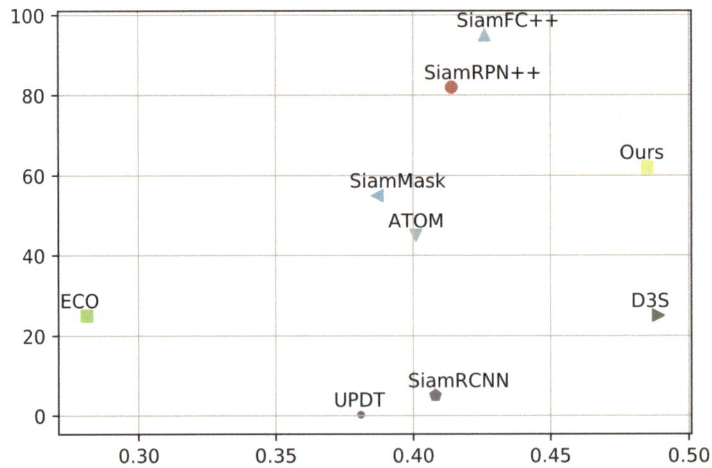

Figure 11.
A comparison of the quality and the speed of state-of-the-art tracking methods on VOT2018.

6. Conclusions

In this work, we propose a deformable correlation network that models geometric transformation from different levels in a highly efficient manner. The comprehensive experiments demonstrate that our approach significantly improves the tracking results, achieving new state-of-the-art performance.

Acknowledgements

This work was supported by the NYUAD Center for Artificial Intelligence and Robotics (CAIR), funded by Tamkeen under the NYUAD Research Institute Award CG010.

Author details

Daitao Xing[1*] and Anthony Tzes[2,3]

1 Department of Computer Science and Engineering, New York University, New York, USA

2 Center for Artificial Intelligence and Robotics, New York University Abu Dhabi, Abu Dhabi, UAE

3 Electrical Engineering, New York University Abu Dhabi (NYUAD), Abu Dhabi, UAE

*Address all correspondence to: daitao.xing@nyu.edu

References

[1] Lee K-H, Hwang J-N. On-road pedestrian tracking across multiple driving recorders. IEEE Transactions on Multimedia. New York City, United States: IEEE; 2015;**17**(9):1429-1438

[2] Tang S, Andriluka M, Andres B, Schiele B. Multiple people tracking by lifted multicut and person re-identification. In: Proceedings of the IEEE Conference on Computer Vision and Pattern Recognition. New York City, United States: IEEE; 2017. pp. 3539-3548

[3] Mueller M, Smith N, Ghanem B. A benchmark and simulator for uav tracking. In: European Conference on Computer Vision. New York City, United States: Springer; 2016. pp. 445-461

[4] Wang Q, Zhang L, Bertinetto L, Hu W, Torr PH. Fast online object tracking and segmentation: A unifying approach. In: Proceedings of the IEEE Conference on Computer Vision and Pattern Recognition. New York City, United States: IEEE; 2019

[5] Lukezic A, Matas J, Kristan M. D3s—A discriminative single shot segmentation tracker. In: Proceedings of the IEEE conference on computer vision and pattern recognition. New York City, United States: IEEE; 2020. pp. 7133-7142

[6] Tripathi AS, Danelljan M, Van Gool L, Timofte R. Tracking the known and the unknown by leveraging semantic information. In: The British Machine Vision Conference. Vol. 2. Durham, UK: BMVC; 2019. p. 6

[7] Čehovin L, Leonardis A, Kristan M. Visual object tracking performance measures revisited. IEEE Transactions on Image Processing. New York City, United States: IEEE; 2016;**25**(3):1261-1274

[8] Kristan M, Leonardis A, Matas J, Felsberg M, Pflugfelder R, Zajc LC, et al. The sixth visual object tracking vot 2018 challenge results. In: Proceedings of the European conference on computer vision workshops. New York City, United States: Springer; 2018. pp. 0-0

[9] Wu Y, Lim J, Yang M-H. Online object tracking: A benchmark. In: Proceedings of the IEEE Conference on Computer Vision and Pattern Recognition. New York City, United States: IEEE; 2013. pp. 2411-2418

[10] Muller M, Bibi A, Giancola S, Alsubaihi S, Ghanem B. Trackingnet: A large-scale dataset and benchmark for object tracking in the wild. In: Proceedings of the European Conference on Computer Vision (ECCV). New York City, United States: Springer; 2018. pp. 300-317

[11] Huang L, Zhao X, Huang K. Got-10k: A large high-diversity benchmark for generic object tracking in the wild. In: IEEE Transactions on Pattern Analysis and Machine Intelligence. Vol. 43. no. 5. New York City, United States: Springer; 2019. pp. 1562-1577

[12] Fan H, Lin L, Yang F, Chu P, Deng G, Yu S, et al. Lasot: A high-quality benchmark for large-scale single object tracking. In: Proceedings of the IEEE Conference on Computer Vision and Pattern Recognition. New York City, United States: IEEE; 2019. pp. 5374-5383

[13] Perazzi F, Pont-Tuset J, McWilliams B, Van Gool L, Gross M, Sorkine-Hornung A. A benchmark dataset and evaluation methodology for video object segmentation. In: Proceedings of the IEEE Conference on Computer Vision and Pattern

Recognition. New York City, United States: IEEE; 2016. pp. 724-732

[14] Pont-Tuset J, Perazzi F, Caelles S, Arbeláez P, Sorkine-Hornung A, Van Gool L. The 2017 davis challenge on video object segmentation. arXiv preprint arXiv:1704.00675. 2017

[15] Li B, Yan J, Wu W, Zhu Z, Hu X. High performance visual tracking with siamese region proposal network. In: Proceedings of the IEEE Conference on Computer Vision and Pattern Recognition. New York City, United States: IEEE; 2018. pp. 8971-8980

[16] Zhu Z, Wang Q, Li B, Wu W, Yan J, Hu W. Distractor-aware siamese networks for visual object tracking. In: Proceedings of the European Conference on Computer Vision (ECCV). New York City, United States: Springer; 2018. pp. 101-117

[17] Li B, Wu W, Wang Q, Zhang F, Xing J, Yan J. Siamrpn++: Evolution of siamese visual tracking with very deep networks. In: Proceedings of the IEEE Conference on Computer Vision and Pattern Recognition. New York City, United States: IEEE; 2019. pp. 4282-4291

[18] Xu Y, Wang Z, Li Z, Yuan Y, Yu G. Siamfc++: Towards robust and accurate visual tracking with target estimation guidelines. In: Association for the Advancement of Artificial Intelligence. Washington DC, United States: AAAI; 2020. pp. 12549-12556

[19] Zhang Z, Peng H, Fu J, Li B, Hu W. Ocean: Object-aware anchor-free tracking. In: The IEEE Conference on European Conference on Computer Vision (ECCV). New York City, United States: Springer; 2020. pp. 771-787

[20] Chen Z, Zhong B, Li G, Zhang S, Ji R. Siamese box adaptive network for visual tracking. In: Proceedings of the IEEE/CVF Conference on Computer Vision and Pattern Recognition. New York City, United States: IEEE; 2020. pp. 6668-6677

[21] Bhat G, Danelljan M, Gool LV, Timofte R. Learning discriminative model prediction for tracking. In: Proceedings of the IEEE International Conference on Computer Vision. New York City, United States: IEEE; 2019. pp. 6182-6191

[22] Danelljan M, Bhat G, Khan FS, Felsberg M. Atom: Accurate tracking by overlap maximization. In: Proceedings of the IEEE Conference on Computer Vision and Pattern Recognition. New York City, United States: IEEE; 2019. pp. 4660-4669

[23] Voigtlaender P, Leibe B. Online adaptation of convolutional neural networks for video object segmentation. arXiv preprint arXiv:1706.09364. 2017

[24] Cheng J, Tsai Y-H, Hung W-C, Wang S, Yang M-H. Fast and accurate online video object segmentation via tracking parts. In: Proceedings of the IEEE Conference on Computer Vision and Pattern Recognition. New York City, United States: IEEE; 2018. pp. 7415-7424

[25] Caelles S, Maninis K-K, Pont-Tuset J, Leal-Taixé L, Cremers D, Van Gool L. One-shot video object segmentation. In: Proceedings of the IEEE Conference on Computer Vision and Pattern Recognition. New York City, United States: IEEE; 2017. pp. 221-230

[26] Yang L, Wang Y, Xiong X, Yang J, Katsaggelos AK. Efficient video object segmentation via network modulation. In: Proceedings of the IEEE Conference on Computer Vision and Pattern

Recognition. New York City, United States: IEEE; 2018. pp. 6499-6507

[27] Yu Y, Xiong Y, Huang W, Scott MR. Deformable siamese attention networks for visual object tracking. In: Proceedings of the IEEE/CVF Conference on Computer Vision and Pattern Recognition. New York City, United States: IEEE; 2020. pp. 6728-6737

[28] Dai J, Qi H, Xiong Y, Li Y, Zhang G, Hu H, et al. Deformable convolutional networks. In: Proceedings of the IEEE International Conference on Computer Vision. New York City, United States: IEEE; 2017. pp. 764-773

[29] Jiang B, Luo R, Mao J, Xiao T, Jiang Y. Acquisition of localization confidence for accurate object detection. In: Proceedings of the European Conference on Computer Vision (ECCV). New York City, United States: Springer; 2018. pp. 784-799

[30] He K, Gkioxari G, Dollár P, Girshick R. Mask r-cnn. In: Proceedings of the IEEE International Conference on Computer Vision. New York City, United States: IEEE; 2017. pp. 2961-2969

[31] Deng J, Dong W, Socher R, Li L, Li K, Fei-Fei L. Imagenet: A large-scale hierarchical image database. In: 2009 IEEE Conference on Computer Vision and Pattern Recognition. New York City, United States: IEEE; 2009. pp. 248-255

[32] He K, Zhang X, Ren S, Sun J. Deep residual learning for image recognition. In: Proceedings of the IEEE Conference on Computer Vision and Pattern Recognition. New York City, United States: IEEE; 2016. pp. 770-778

[33] Lin T-Y, Maire M, Belongie S, Hays J, Perona P, Ramanan D, et al. Microsoft coco: Common objects in context. In: European Conference on Computer Vision. New York City, United States: Springer; 2014. pp. 740-755

[34] Xu N, Yang L, Fan Y, Yue D, Liang Y, Yang J, et al. Youtube-vos: A large-scale video object segmentation benchmark. arXiv preprint arXiv: 1809.03327. 2018

[35] Long J, Shelhamer E, Darrell T. Fully convolutional networks for semantic segmentation. In: Proceedings of the IEEE Conference on Computer Vision and Pattern Recognition. New York City, United State: IEEE; 2015. pp. 3431-3440

[36] Tang M, Gorelick L, Veksler O, Boykov Y. Grabcut in one cut. In: Proceedings of the IEEE International Conference on Computer Vision. New York City, United State: IEEE; 2013. pp. 1769-1776

[37] Kingma DP, Ba J. Adam: A method for stochastic optimization. arXiv preprint arXiv:1412.6980. 2014

[38] Zhang L, Gonzalez-Garcia A, Weijer JVD, Danelljan M, Khan FS. Learning the model update for siamese trackers. In: Proceedings of the IEEE International Conference on Computer Vision. New York City, United State: IEEE; 2019. pp. 4010-4019

[39] Bertinetto L, Valmadre J, Henriques JF, Vedaldi A, Torr PH. Fully-convolutional siamese networks for object tracking. In: European Conference on Computer Vision. New York City, United States: Springer; 2016. pp. 850-865

[40] Danelljan M, Bhat G, Shahbaz Khan F, Felsberg M. Eco: Efficient convolution operators for tracking. In: Proceedings of the IEEE Conference on Computer Vision and Pattern Recognition. New York City, United State: IEEE; 2017. pp. 6638-6646

[41] Lukezic A, Vojir T, Zajc LC, Matas J, Kristan M. Discriminative correlation filter with channel and spatial reliability. In: Proceedings of the IEEE Conference on Computer Vision and Pattern Recognition. New York City, United State: IEEE; 2017. pp. 6309-6318

[42] Danelljan M, Robinson A, Khan FS, Felsberg M. Beyond correlation filters: Learning continuous convolution operators for visual tracking. In: European Conference on Computer Vision. New York City, United States: Springer; 2016. pp. 472-488

[43] Dai K, Wang D, Lu H, Sun C, Li J. Visual tracking via adaptive spatially-regularized correlation filters. In: Proceedings of the IEEE Conference on Computer Vision and Pattern Recognition. New York City, United State: IEEE; 2019. pp. 4670-4679

[44] Yang T, Xu P, Hu R, Chai H, Chan AB. ROAM: Recurrently optimizing tracking model. In: Proceedings of the IEEE Conference on Computer Vision and Pattern Recognition. New York City, United State: IEEE; 2020

[45] Du F, Liu P, Zhao W, Tang X. Correlation-guided attention for corner detection based visual tracking. In: Proceedings of the IEEE/CVF Conference on Computer Vision and Pattern Recognition. New York City, United State: IEEE; 2020. pp. 6836-6845

[46] Nam H, Han B. Learning multi-domain convolutional neural networks for visual tracking. In: Proceedings of the IEEE conference on computer vision and pattern recognition. New York City, United State: IEEE; 2016. pp. 4293-4302

[47] Voigtlaender P, Luiten J, Torr PH, Leibe B. Siam r-cnn: Visual tracking by re-detection. In: Proceedings of the IEEE/CVF Conference on Computer Vision and Pattern Recognition. New York City, United State: IEEE; 2020. pp. 6578-6588

[48] Danelljan M, Gool LV, Timofte R. Probabilistic regression for visual tracking. In: Proceedings of the IEEE/CVF Conference on Computer Vision and Pattern Recognition. New York City, United State: IEEE; 2020. pp. 7183-7192

[49] Bhat G, Johnander J, Danelljan M, Shahbaz Khan F, Felsberg M. Unveiling the power of deep tracking. In: Proceedings of the European Conference on Computer Vision (ECCV). New York City, United States: Springer; 2018. pp. 483-498

Chapter 5

Modeling and Simulation of a Solar-Powered Quadcopter Using MATLAB

Rahmi Elagib and Ahmet Karaarslan

Abstract

Solar-powered quadcopters have the potential to revolutionize the way we think about aerial transportation and surveillance. These aircraft can potentially fly for extended periods of time without the need for external fuel, making them ideal for a wide range of applications. In this paper, we present a modeling and simulation approach for a solar-powered quadcopter using MATLAB. The model includes the quadcopter's dynamics, solar panel power generation, and energy storage system. A PID control system for the solar-powered quadcopter simulation model was created using MATLAB/Simulink. The simulation findings indicate that the solar-powered quadcopter can be accurately simulated and controlled with a PID controller. It also reveals that a solar-powered quadcopter can fly for 2.18 hours with a state of charge (SOC) of more than 20%, compared to a nonsolar quadcopter's flying duration of 34.37 minutes. The results of this study can be used to design and optimize the performance of solar-powered quadcopters for various applications.

Keywords: quadcopter, MATLAB, PID, solar-powered, unmanned aerial vehicles

1. Introduction

Solar-powered quadcopters have gained significant attention in recent years due to their potential for extended flight and their ability to operate in a wide range of environments. These aircraft are typically equipped with photovoltaic cells, which convert sunlight into electricity that can be used to power the quadcopter's motors and electronics. However, there are many challenges to overcome in order to make solar-powered quadcopters a practical reality, including the design of efficient and lightweight solar panels, the optimization of the energy storage system, and the management of the quadcopter's power consumption.

Modeling and simulation can be a powerful tool for understanding the performance and behavior of solar-powered quadcopters. By building a virtual model of the quadcopter and its components, it is possible to study the effects of various design choices and operating conditions without the need for physical experimentation. In this paper, we present a modeling and simulation approach for a solar-powered quadcopter using MATLAB. The model includes the quadcopter's dynamics, solar panel power generation, and energy storage system. The simulation results are used to study the performance of the solar-powered quadcopter under different solar insolation conditions.

There have been several previous studies on the use of solar power for small unmanned aerial vehicles (UAVs), including solar-powered quadcopters. These studies have primarily focused on the use of photovoltaic cells to generate electricity for the UAV's motors and electronics. However, there have been few comprehensive studies on the design and performance of a fully solar-powered quadcopter.

Many researchers proposed solar-powered quadcopters that use hybrid energy sources to increase flight durations, such as integrating rotational energy harvesting, laser power beaming, and solar energy [1–3] J. Meyer et al. [4] present a design for sustained solar-powered electric flight on small scale Low Altitude Long Endurance (LALE) UAVs. This model improved the airframe construction, energy storage mediums, motor and propeller efficiency, and the availability of lightweight photovoltaic cells for UAV power generation. Goh et al. [5] used 148 solar cells to create a fully solar-powered quadcopter. It weighed about 2.6 kg and reached a height of 10 m. Even after building a massive working model, their quadcopter's overall flight time was only about 1 minute and 38 seconds. Kingry et al. [6] developed a solar-powered quadcopter using an array of C60 solar cells with 22% efficiency. Because they wanted to attain a very high-flying time, which requires using numerous solar cells, their demonstrator weighed roughly 8.175 kg. The flight time of this solar-powered quadcopter was approximately an hour. Verbeke et al. evidenced a modified configuration for narrow corridors that results in up to 60% more endurance [7]. Pang et al. recently incorporated variable pitch rotors into a gasoline engine to extend flight duration to 2 to 3 hours [8].

Despite the development of small-scale solar-powered quadcopters in [9–12], there aren't many studies that address the modeling and control of a solar-powered quadcopter problem. Proportional-integral-derivative (PID) control, model predictive control, and PID-LQR integrated control are examples of effective control systems that have been used to develop robust and effective controllers in various solar-powered quadcopters. In this study, a 1.147 kg solar-powered quadcopter will be modeled, simulated, and PID-controlled in MATLAB/Simulink. The main contribution of this study is the building of a realistic MATLAB/Simulink model for a solar-powered quadcopter in order to increase flying duration while taking into consideration the issue of controlling and stabilizing the vehicle, which can serve as a solid basis for designing an actual model.

This work is divided into the following sections: Section II includes the Model Description, and Section III depicts the PID Controller Design for a solar-powered quadcopter. Section IV discusses the Simulation Results. Section V covers the Conclusion as well as future work.

2. Model description

The model of the solar-powered quadcopter consists of three main components: the quadcopter dynamics, the solar panel power generation, and the energy storage system.

2.1 The quadcopter dynamics model

In this section, a mathematical model of quadcopter flight dynamics was developed. The quadcopter's linear position is defined as (x, y, z), and the three Euler angles are defined as (ϕ, θ, ψ). The quadcopter has four motors, two of which spin clockwise and the others counterclockwise, so the torque generated is negated.

A quadcopter's altitude and attitude can be controlled by varying the speed of each rotor $(\Omega_1, \Omega_2, \Omega_3, \Omega_4)$.

The flight dynamics model of the solar-powered quadcopter is built on a number of assumptions, such as the quadcopter's rigid and symmetrical structure, the coincidence of the center of gravity and the origin of the body fixed frame, and proportional relationships between thrust/drag torques and the square of rotor speed, and rigid propellers. The flight dynamics model of a quadcopter is derived from Newton-Euler equations and Newton's second law as follows:

$$\ddot{x} = \frac{1}{m}((\sin\psi\sin\phi - \cos\psi\sin\theta\sin\phi)U_1 - A_x\dot{x}) \tag{1}$$

$$\ddot{y} = \frac{1}{m}((\cos\psi\sin\phi + \sin\psi\sin\theta\cos\phi)U_1 - A_y\dot{y}) \tag{2}$$

$$\ddot{z} = g - \frac{1}{m}(\cos\psi\cos\phi U_1 - A_y\dot{y}) \tag{3}$$

$$\ddot{\phi} = \left(\frac{I_{yy} - I_{zz}}{I_{xx}}\right)\dot{\psi}\dot{\theta} + \left(\frac{J_r\Omega_r}{I_{xx}}\right)\dot{\theta} + \left(\frac{1}{I_{xx}}\right)U_2 - \frac{A_\phi\dot{\phi}}{I_{xx}} \tag{4}$$

$$\ddot{\theta} = \left(\frac{I_{zz} - I_{xx}}{I_{yy}}\right)\dot{\psi}\dot{\phi} - \left(\frac{J_r\Omega_r}{I_{yy}}\right)\dot{\phi} + \left(\frac{1}{I_{yy}}\right)U_3 - \frac{A_\theta\dot{\theta}}{I_{yy}} \tag{5}$$

$$\ddot{\psi} = \left(\frac{I_{xx} - I_{yy}}{I_{zz}}\right)\dot{\theta}\dot{\phi} + \left(\frac{1}{I_{zz}}\right)U_4 - \frac{A_\psi\dot{\psi}}{I_{zz}} \tag{6}$$

The input signal U_1 is the total thrust of the four rotors. And U_2, U_3, U_4 are the moments for pitch, roll, and yaw, respectively. Where m represents the mass of the quadrotor, J_r is the inertia of the rotor, and I_{xx}, I_{yy}, and I_{zz} are the inertia of the quadrotor in x, y, and z, respectively. More information can be found about the dynamics model in our previous works [13, 14].

2.2 The solar power system modeling and simulation

Instead of using just solar energy, a hybrid system of LiPo battery and solar energy is designed to extend the entire flying duration of the quadrotor. This system consists mostly of PV arrays, maximum power point tracking (MPPT), a buck converter, and a battery. Because the modeling of such systems has been extensively studied in several studies, our primary focus here will be the simulation of this system.

Maximum power point tracking (MPPT) is a solar charge controller that controls the amount of power from the solar array feeding the battery. It prevents electricity from running back to the solar panels overnight and prevents the deep cycle batteries from being overloaded during the day. A DC-to-DC transformer, the MPPT charge controller can convert power from a higher voltage to power at a lower voltage. Since the quantity of power remains constant, if the output voltage is less than the input voltage, the output current will be greater than the input current, maintaining the constant value of the product P=VI. Therefore, a charge controller should be able to select the ideal current-voltage point on the current-voltage curve to get the most out

Parameter	Symbol	Value
Maximum power	P_{mpp}	3.42 W
Voltage at maximum power	V_{mpp}	0.582 V
Current at maximum power	I_{mpp}	5.93A
Open-circuit voltage	V_{OC}	0.687 V
Short-circuit Current	I_{SC}	6.28I
Cell efficiency	η	22.5%

Table 1.
SunPower C60 cell electrical characteristics.

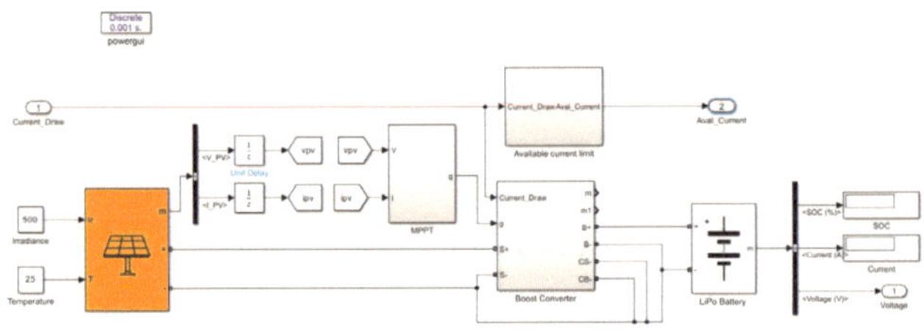

Figure 1.
The Simulink of built solar-powered quadcopter battery circuit.

of a solar panel. Buck converter, the MPPT is built on a synchronous buck converter circuit. It reduces the greater solar panel voltage to the battery's charging voltage. Additionally, it modifies the voltage of its input to capture the solar panel's maximum output before transforming that output to meet the battery's fluctuating voltage needs. Solar panels, the quadcopter's solar power system's key component is the solar panel. There are several names for solar panels, including photovoltaic solar modules, solar plates, solar PV modules, and solar power panels. PV arrays are made up of 30 solar cells that create power to charge the battery. The most important consideration when choosing a solar cell for a quadcopter is efficiency. As a result, it was decided to employ SunPower C60 solar cells, irradiation 500 W/m^2, 25C. Under conventional testing settings, each cell possesses the electrical characteristics as shown in **Table 1** (STC). PV array voltage can be higher than battery voltage for MPPT. Lithium-ion polymer (LiPo), the battery utilized in simulation is a 3200 mAh Li-Po battery with configuration: 3S1P/11.1v/3Cell, constant discharge: 30C, and peak discharge (10 sec): 60C. The simulation of the proposed solar-powered quadcopter power system (solar cells, a Li-Po battery, an MPPT, and a boost converter) is shown in **Figure 1**.

3. Quadcopter PID controller

PID is used in this investigation to obtain the desired altitude and attitude. The proposed control law is developed by dividing the system model into two subsystems,

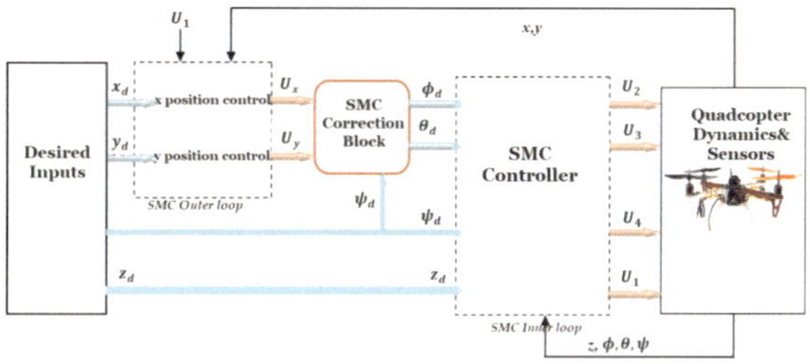

Figure 2.
UAV control system block diagram.

Parameter	Symbol	Value
Quad. mass	m	1.147 kg
Arm length	l	0.36 m
Gravity	g	9.81 m/s^2
Rotor inertia	J_r	2.6e-06 $kg.m^2$
Inertia constants	I_{yy}, I_{xx}	$9.5^* \ 10^{-3}$ $kg.m^2$
	I_{zz}	$1.86^* \ 10^{-2}$ $kg.m^2$
Thrust coeff.	b	1.4865e-07 $N.s^2$
Drag coeffi.	d	2.925e-09 $N.m.s^2$

Table 2.
The parameters of the quadcopter.

a fully actuated subsystem and an under-actuated subsystem, as shown in **Figure 2**. Unlike in the under-actuated subsystem, where the inputs U_2 and U_3 are smaller than the number of outputs (x, y, ϕ, θ), in the fully actuated subsystem, there are two outputs (z, ψ) for each of the inputs (U_1, U_4).

Many articles, including [15], have investigated the PID controller. The PID controller has three different parameters: the proportional term, the integral term, and the derivative term. The proportional term determines the direct action with regard to the computed error, the integral term determines the response with respect to the sum of recent mistakes, and the derivative term determines the reaction with respect to the error's rate of change. The controller equation is given by the following formula:

$$U_X(t) = K_p e(t) + K_I \int e(t).dt + K_D \frac{de(t)}{dt} \tag{7}$$

In Simulink, PID controllers may be represented as gains or as a continuous time transfer function (**Table 2**).

4. Simulation results

The suggested PID design parameters have been tuned manually in MATLAB/Simulink in order to track the trajectory smoothly. **Table 3** lists the parameters of the recommended controllers.

Controller Tuning	PID			
	Roll	Pitch	Yaw	Altitude
K_P	1.99	1.99	0.12	1.8
K_I	1.99	1.99	0.12	2.1
K_D	0.11	0.11	0	1.9

Table 3.
PID design parameters tuning.

Figures 3–6 show the solar-powered quadcopter's response to altitude, roll, pitch, and yaw velocity. The system response is excellent, and the control procedure is carried out with very minor overshoots. However, there is a noticeable overshoot in the case of altitude and yaw velocity control, at 60 seconds for altitude and 66, and 71 seconds for yaw velocity.

A trajectory can be simulated to show that the simulation is appropriately controlled and operating. A 3D trajectory may be created using the trajectory inputs and states of altitude, roll, pitch, and yaw velocity. **Figure 7** shows the three-dimensional trajectory, whereas **Figure 8** illustrates the sensor position output.

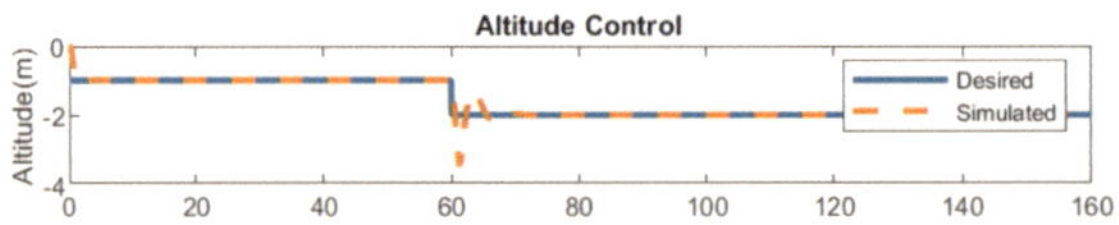

Figure 3.
The actual and desired altitude values.

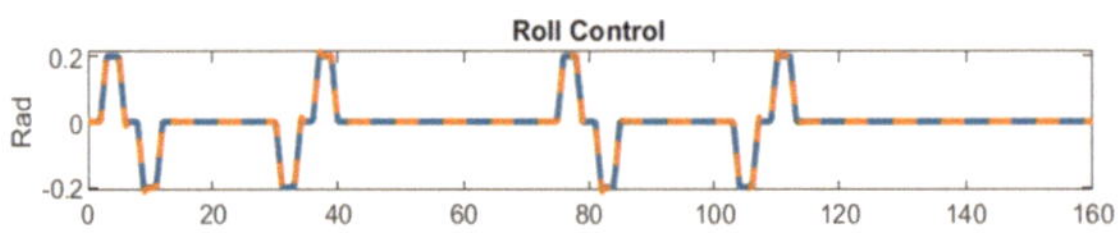

Figure 4.
The actual and desired roll values.

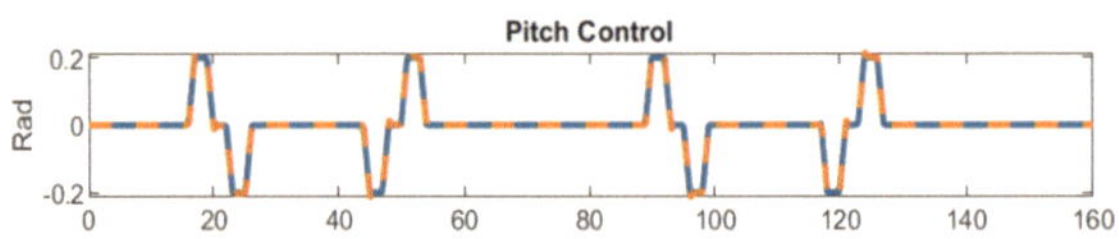

Figure 5.
The actual and desired pitch values.

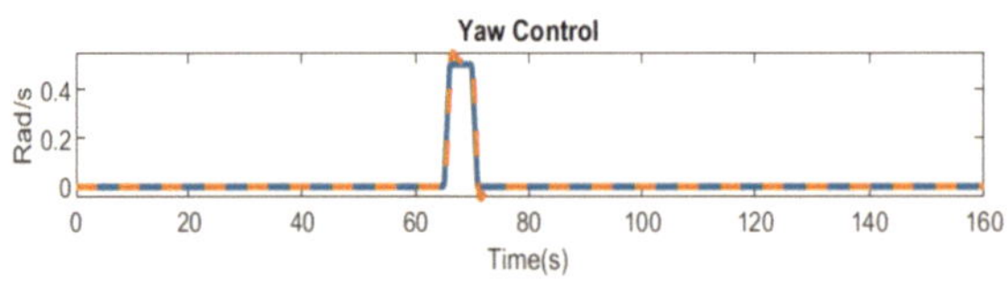

Figure 6.
The actual and desired yaw values.

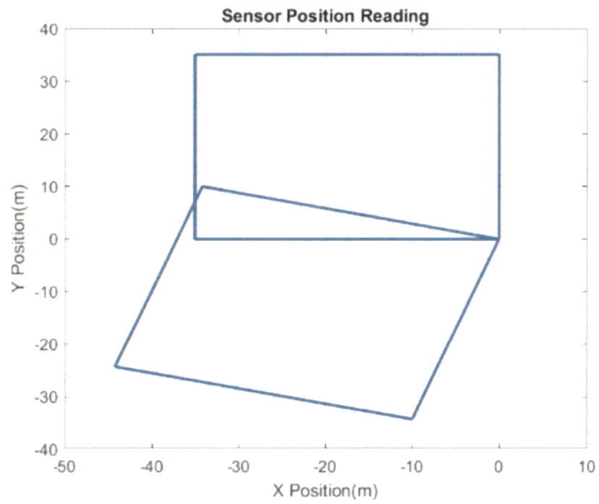

Figure 7.
The quadcopter position sensor reading.

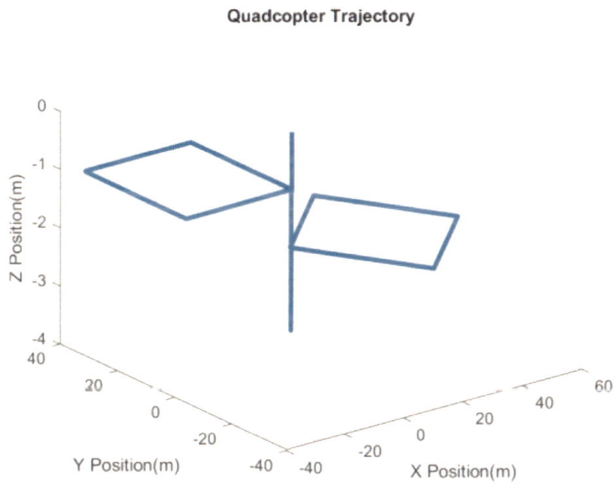

Figure 8.
The quadcopter trajectory.

The RPM of the motors is also provided in **Figure 7** for further controller demonstration. The aforementioned overshoot may also be seen in the RPM diagram at the same time period (**Figure 9**).

The quadcopter's voltage, current, and battery state of charge (SOC) are simulated in both cases, with and without solar cells. The results are compared to show how using solar cells improves flight duration. **Figures 10–17** compare battery simulation results with and without solar cells.

Figures 10 and **11** illustrate that when solar cells are used, the voltage is higher than when they are not used.

Figures 12 and **13** show that when solar cells are employed, the current drawn is lower than when they are not. When determining the quadcopter flying time, these current values are critical.

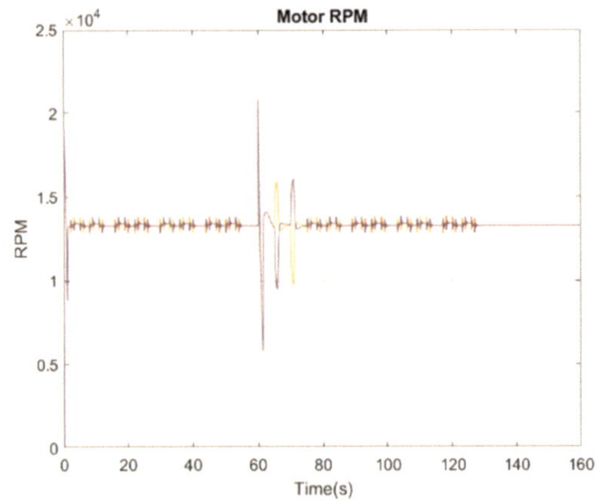

Figure 9.
The quadcopter RPM.

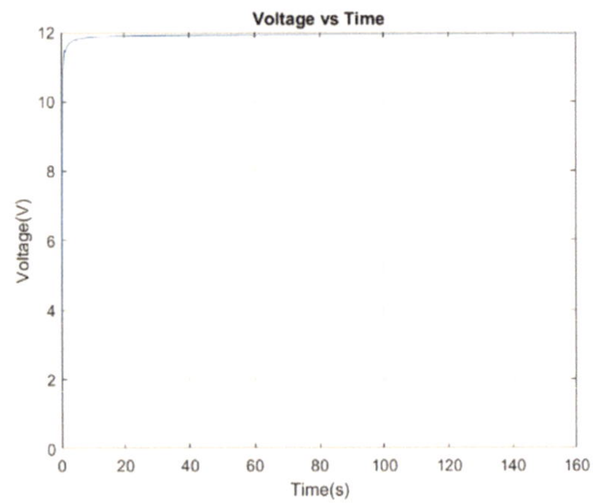

Figure 10.
The battery voltage vs. time when solar cells are used.

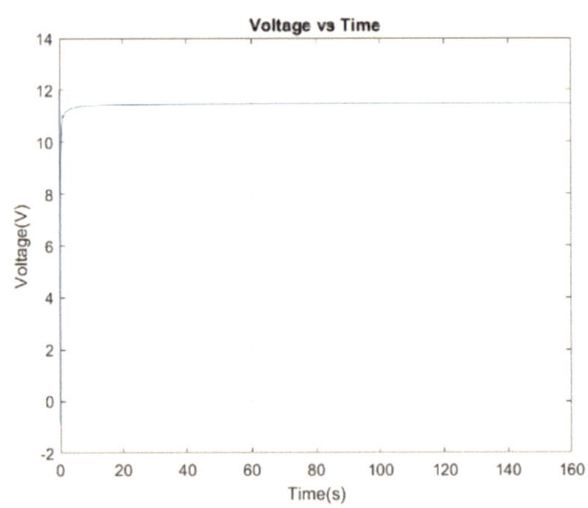

Figure 11.
The battery voltage vs. time when no solar cells are used.

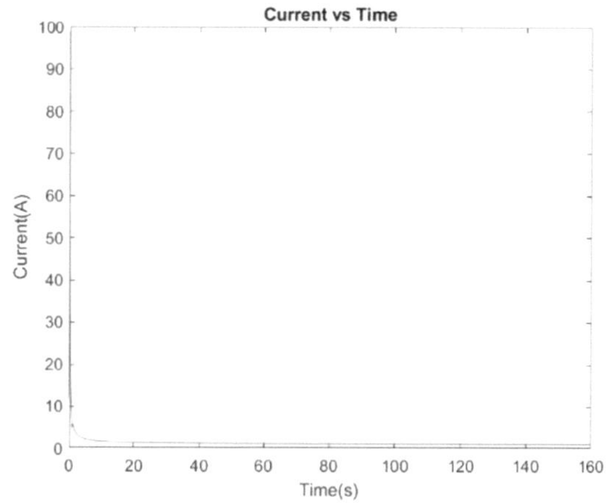

Figure 12.
The battery current vs. time when solar cells are used.

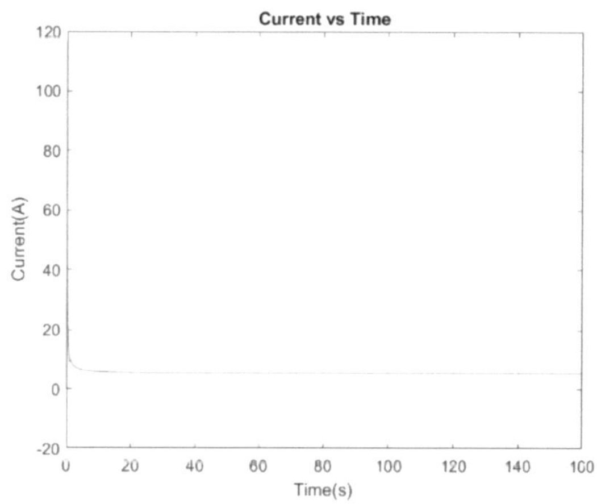

Figure 13.
The battery current vs. time when no solar cells are used.

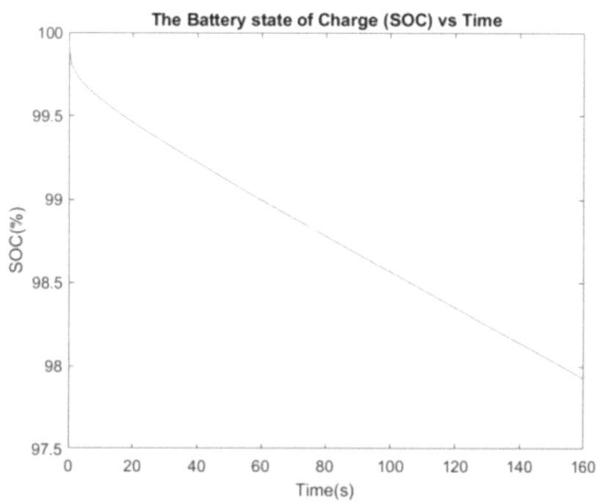

Figure 14.
The battery SOC vs. time when solar cells are used.

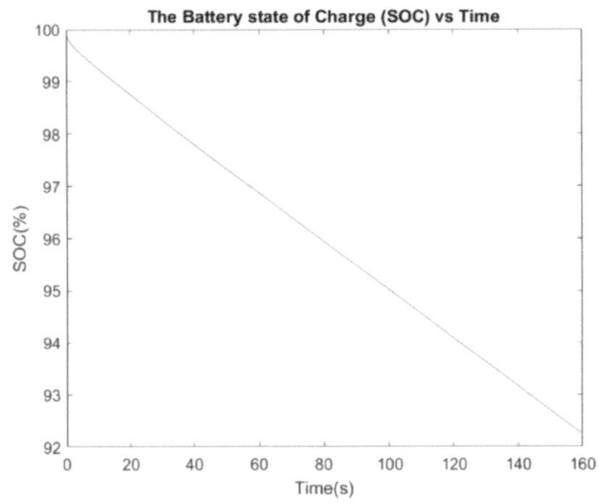

Figure 15.
The battery SOC vs. time when no solar cells are used.

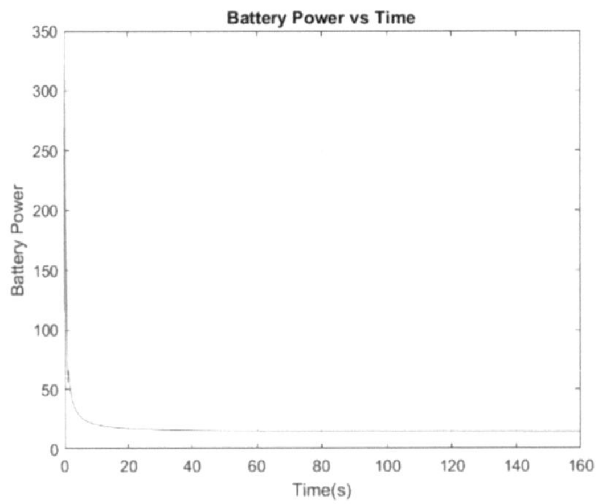

Figure 16.
The battery power vs. time when solar cells are used.

Figures 14 and **15** show that the SOC is lower when solar cells are employed than when they are not used. This means that when solar cells are not employed, the quadcopter discharges more quickly.

Figures 16 and **17** show that when solar cells are utilized, the battery power gradually decreases because the quadcopter uses a part of the cell current, causing the battery power to decrease due to the drop in battery current.

The flight time can be calculated in both situations using the following formula:

$$\text{The quadcopter flight time} = \text{The battery capacity} \times \text{save discharge rate}/\text{AAD} \quad (8)$$

$$\text{AAD} = \text{AUW} \times \text{P}/\text{V} = \text{AUW} \times \text{I} \quad (9)$$

where P is the power needed to lift one kilogram of equipment, expressed in watts per kilogram, I is the current (in amps) needed to lift one kilogram into the air, and AAD is the average amp draw, expressed in amperes. AUW is the all-up weight of your drone, which is the total weight of the equipment that goes up in the air,

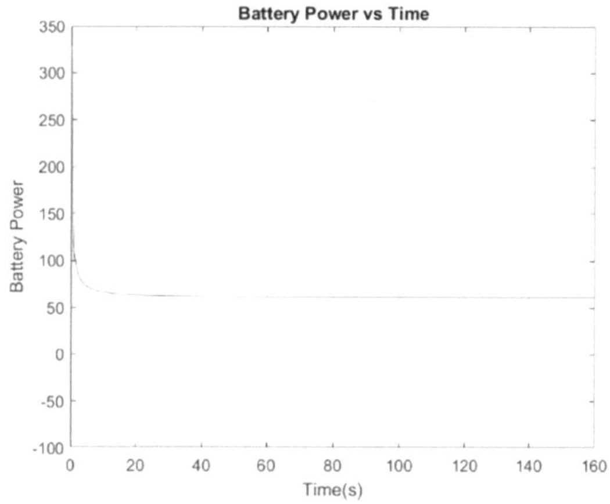

Figure 17.
The battery power vs. time when no solar cells are used.

including the battery. Also, it can be calculated directly from Omni calculator website [16] or using trigonometry.

Flight times are calculated using Eqs. (8) and (9).

Flight time is 34.37 minutes when no solar cells are used, and 2.18 hours when solar cells are used.

5. Conclusion

This article discusses the modeling and control of a solar-powered quadcopter, resulting in significantly longer flight duration. The quadcopter's flight dynamics and aerodynamics models are being developed. To simulate the model in MATLAB, a mathematical model in Newton-Euler form was derived. An accurate plant model was built using MATLAB/Simulink, and motor dynamics and battery dynamics were integrated. The solar-powered quadcopter power system was modeled utilizing SunPower C60 sun cells and an MPPT, Buck Converter. These parameters were added to the plant model to make it more realistic for real-world application. The designed solar-powered vehicle was controlled by PID controllers, and the system reaction was recorded after the PID controllers had been tuned. The simulation results demonstrate that the system is properly controlled and operated.

Furthermore, the results reveal that the solar-powered quadcopter performs well under typical solar irradiation circumstances. While the nonsolar-powered quadcopter can fly for 34.3 minutes, the solar-powered quadcopter can fly for 2.18 hours with the battery's state of charge (SOC) maintaining over 20%. Although the angle of the panels is greatly important on solar panel power output, this article does not focus on this issue; thus, additional research on this topic may be done in future works. A real model will be developed and tested outdoors in the future to compare simulation results with real-world results.

Acknowledgements

I'd like to thank Prof. Dr. Ahmet Karaarslan for his patient guidance, support, and counsel.

Conflict of interest statement

The authors have no conflicts of interest to declare that are relevant to the content of this article.

Author details

Rahmi Elagib* and Ahmet Karaarslan
Department of Electrical and Electronics Engineering, Ankara Yildirim Beyazıt University, Ankara, Turkey

*Address all correspondence to: rahmielagib@gmail.com; 185105116@ybu.edu.tr

References

[1] Sowah R, Acquah MA, Ofoli AR, Mills GA, Koumadi KM. Rotational energy harvesting to prolong flight duration of quadcopters. In: 2015 IEEE Industry Applications Society Annual Meeting. Addison, TX, USA; 2015. pp. 1-7. DOI: 10.1109/IAS.2015.7356835

[2] Shaheed MH et al. Flying by the sun only: The solarcopter prototype. Aerospace Science and Technology. 2015;**45**:209-214. DOI: 10.1016/j.ast.2015.05.016

[3] Achtelik MC, Stumpf J, Gurdan D, Doth K-M. Design of a flexible high performance quadcopter platform breaking the MAV endurance record with laser power beaming. In: 2011 IEEE/RSJ International Conference on Intelligent Robots and Systems. San Francisco, CA, USA; 2011. pp. 5166-5172. DOI: 10.1109/iros.2011.6094731

[4] African Electrical Technology Conference Institute of Electrical and Electronics Engineers South African Section & Polytechnic of Namibia. (2007). IEEE; 2007. Available from: http://ieeexplore.ieee.org/servlet/opac?punumber=4401437 [Retrieved: September 25, 2023]

[5] Goh CS, Kuan JR, Yeo JH, Teo BS, Danner A. A fully solar-powered quadcopter able to achieve controlled flight out of the ground effect. Progress in Photovoltaics: Research and Applications. 2019;**27**(10):869-878. DOI: 10.1002/pip.3169

[6] Kingry N, et al. Design, Modeling and Control of a Solar-Powered Quadcopter. In: 2018 IEEE International Conference on Robotics and Automation (ICRA). Brisbane, QLD, Australia; 2018. pp. 1251-1258. DOI: 10.1109/ICRA.2018.8462896

[7] Verbeke J, Hulens D, Ramon H, Goedemé T, De Schutter J. The design and construction of a high endurance hexacopter suited for narrow corridors. In: 2014 International Conference on Unmanned Aircraft Systems (ICUAS). Orlando, FL, USA. 2014. pp. 543-551. DOI: 10.1109/ICUAS.2014.6842296

[8] Pang T, Peng K, Lin F, Chen BM. Towards long-endurance flight: Design and implementation of a variable-pitch gasoline-engine quadrotor. In: 2016 12th IEEE International Conference on Control and Automation (ICCA). Kathmandu, Nepal; 2016. pp. 767-772. DOI: 10.1109/ICCA.2016.7505371

[9] Bemporad A, Pascucci CA, Rocchi C. Hierarchical and hybrid model predictive control of quadcopter air vehicles. In: IFAC Conference on Analysis and Design of Hybrid Systems. 2009

[10] IEEE Robotics and Automation Society. ICRA 2009: 2009 IEEE International Conference on Robotics and Automation, Kobe, Japan. IEEE Press; 2009. ISBN: 978-1-4244-2788-8

[11] Tran NK, Bulka E, Nahon M. Quadrotor control in a wind field. In: 2015 International Conference on Unmanned Aircraft Systems (ICUAS), Denver, CO, USA. 2015. pp. 320-328. DOI: 10.1109/ICUAS.2015.7152306

[12] Amin R, Aijun L, Shamshirband S. A review of quadrotor UAV: Control methodologies and performance evaluation. International Journal of Automation and Control. 2016;**10**(2):87-103. DOI: 10.1504/IJAAC.2016.076453

[13] Elagib R, Karaarslan A. Stabilization of Quadcopter Using the Combination of

LQR And SMC Methods Conference: International Marmara Sciences Congress, Kocaeli, TÜRKİYE. 2022; **13**(2):283-292. ISBN: 978-605-70762-4-3

[14] Elagib R, Karaarslan A. Implementation and stabilization of a quadcopter using Arduino and the combination of LQR and SMC methods. Journal of Engineering Research and Reports (JERR), London, UK. 2022:42-58. DOI: 10.9734/jerr/ 07 2022/v23i7735

[15] Romero LE, Pozo DF, Rosales JA. Quadcopter Stabilization by Using PID Controllers. Quito, Ecuador: MASKANA; 2014

[16] Omni Calculator. Drone Flight Time Calculator. Available from: https://www.omnicalculator.com/other/drone-flight-time [Accessed: September 26, 2023]

Chapter 6

Implementation of a Hybrid Power Testbed Model of a Hybrid Electric Unmanned Aerial Vehicle

Lionel Fouellefack

Abstract

Testbeds form an essential aspect in the construction of a Hybrid – Electric Unmanned Aerial Vehicle (HE – UAV)/Aircraft. Such testbeds have never been developed before even though a model of the HE – UAV was developed. This article explores the feasibility of implementing a prototype of a (HE – UAV) testbed. Several research papers in the domain were thoroughly explored and the results were grouped based on similar research findings. The grouping was done in terms of the sizing result achieved by each author, the UAV class used, the hybrid powertrain specifications of the components, and the testbed equipment used to construct the HE – UAV testbed. The result from this research shows that a HE – UAV testbed can be achieved if stringent measures are done in determining the size of the component for a given drone size and a careful selection of the components from the sizing result for the testbed construction.

Keywords: unmanned aerial vehicle, unmanned aerial system, testbed, Hybrid – Electric, aircraft

1. Introduction

A MATLAB – Simulink model of a Hybrid – Electric Unmanned Aerial Vehicle (HE – UAV) was implemented in the previous research to analyze the fuel savings potential of the UAV for specific mission profiles Fouellefack et al. [1]. This research article explores the feasibility of constructing a prototype of a HE – UAV. Similar research studies have already been done by previous authors in the field, which are: sizing the components of the hybrid powertrain of the HE-UAV for specific mission profiles, developing a numerical simulation model through a control strategy to optimize HE-UAV performance and constructing test beds to integrate the components of hybrid powertrain which formed the basis for the build-up of a flying UAV/Aircraft. These are explored in the relevant sections below.

1.1 Component sizing

One Aspect essential in the construction of a HE-UAV is to do component sizing of the hybrid powertrain components of the UAV/Aircraft. By component sizing, the author refers to the specification of the components of the hybrid powertrain used to achieve a mission for a given UAV design. Several research studies have been attempted by previous researchers [2–8] to determine the size of each propulsive component of a hybrid powertrain as this is essential for the testbed construction, UAV size and range covered by the UAV. These include developing mythologies through various optimization functions to minimize specific design parameters to reduce fuel consumption and increase the UAV range and implementing multiple sizing functions to estimate the specification of each propulsive component to achieve a given UAV mission.

Using a non-dominated sorting genetic algorithm, Xie et al. [2] developed a sizing algorithm to size the Hybrid Electric Propulsion Systems (HEPS) for a small UAV for specific mission profiles. The algorithm developed considers two objectives; fuel consumption and flight during the UAV to estimate its range and fuel consumption which was compared to a prototype aircraft. The hybrid aircraft achieved a range of 6 h compared to 5 h using the prototype aircraft resulting in a fuel economy of 17.6%. A similar study was done by Donateo et al. [3] Donateo et al. [4], where the authors developed a multi-objective optimization problem using a genetic algorithm to size the components of a HE – UAV for specific mission profiles. He developed scalable models to calculate the mass and space of each powertrain component (Propeller, gearbox, engine, battery and electric drive) and an optimization routine was run to maximize the endurance and minimize the fuel consumption of the UAV during its mission profile. The optimization of the mission allowed an improvement in fuel consumption of 7% compared to the existing configuration. Pornet et al. [5] produced a similar performance methodology for sizing hybrid electric aircraft for specific mission profiles. His sizing algorithm was used to size the plane for some design variables, and an objective function was developed to minimize fuel consumption during the aircraft's mission. A fuel consumption of 16% was achieved compared to the conventional aircraft with a mix of fuel and electrical energy of 82%:18%.

Hiserote et al. [6] developed a sizing algorithm using MATLAB to size the propulsion system component of a HE – UAV. The code was designed such that it takes inputs such as the user-specific mission, i.e. airspeed, altitude, rate of climb, battery discharge and hybrid configuration to optimize the power required by the aircraft and perform various sizing calculations of the propulsion system, such as the ICE, EM mass, propeller mass, clutch mass, and battery mass. The author used a 3 h ISR baseline mission profile and compared the size of the propulsion system with the conventional case (ICE-powered) UAV. The hybrid UAV achieved a lesser fuel saving than the conventional UAV.

Schomann et al. [7] designed a sizing methodology to size the propulsion system of a HE-UAV. He developed surrogate models of the components of the hybrid propulsion system, i.e., the ICE, EM, Battery, Propeller and fuel system, such that it iteratively outputs specific parameters for a given input. A similar work is seen in Rippel [8] from the Air Force Institute of Technology, where he developed a sizing algorithm for the conceptual design of a HE-UAV. The sizing code iteratively outputs specific design parameters, such as the power required for flight, wingspan, wing area and propulsion system mass (ICE, EM, propeller mass and battery) for a given user input, such as the aircraft altitude, cruise condition and rate of climb.

Table 1 provides a comparative summary of the drone class used for the sizing calculations, **Table 2** gives a comparative overview of the range, endurance and percentage of fuel improvement compared to the conventional case (ICE powered aircraft/UAV) obtained by different authors from their sizing results, and **Table 3**

Paper	Reference UAV/Aircraft class for sizing	Take-off mass (kg)	Wing span (m)
Xie [2]	ERCO Ercoupe 415	603	9.1
Donateo [3]	General Atomics Predator RQ	1020	14.8
Pornet [5]	Airbus A320 or Boeing B737	78,990	36.0
Hiserote [6]	Insitu's ScanEagle AAI's Aerosonde	13.6	4.62
Schomann [7]	Institution unmanned research aircraft IMPULLS	35.49	5.23
Rippel [8]	Diamond Aircraft DA 20	873	10.9
	Cessna 172 Skyhawk,	1114	11
	General Atomics Predator	1190	11.5
Harmon [9]	Aerosonde UAV	13.6	4.65

Table 1.
Comparison of various drone classes for sizing calculation used by each author.

Mission altitude (m)	Cruising velocity	Range	Endurance (h)	Percentage of fuel improvement
230 m	157 km/h	600 km	6	17.6
4900 m	41	—	24	7
35,000 ft	250 K CAS	2770 nm	1.2	16
300 m	20.5 m/s	—	12	30.5
900 m	20	—	12	—
4000	71	547 nm	—	—
4115	60	700 nm	—	—
4500	47	2000 nm	—	—
(1524) 5000 ft	23.2–28.3	—	3	38 22

Table 2.
Comparison of range, endurance, and fuel improvement from sizing results of each HE-UAV achieved by various authors.

compares the sizing results of the hybrid electric propulsion characteristics of the HE-UAV achieved by various authors.

1.2 Supervisory control and optimization

Another important aspect essential for the construction of the HE – UAV testbed is do a numerical simulation of a modeled UAV to predict UAV performance under varying mission profiles. This is usually done by employing a control strategy through an energy management strategy to optimize the power demand of the powertrain components of the UAV during its flight profile. Previous research has applied these control strategies as energy management strategies to optimize the fuel consumption of a HE – UAV during its mission profile. Examples include the usage of a rule base supervisory controller by Harmon et al. [9], Xie et al. [10], Hung et al. [11], where

Paper	ICE	EM	Battery	Propeller
Xie [2]	ICE Power (40 kW)	EM Power (30 kW)	Battery Mass (10 kg)	—
Donateo [3]	Engine Power (64 kW)	EM Power (3KW)	—	—
Pornet [5]		EM Power (5.1 MW) EM Mass (510 kg)	Battery Mass (8900 kg)	—
Hiserote [6] Clutch-Start Electric-Start Centerline-Thrust	Engine Power (438.9 W) Engine Mass (0.356 kg) Engine Power (429.9 W) Engine Mass (0.349 kg) Engine Power (406.5 W) Engine Mass (0.330 kg)	Electric Motor Power (155.2 W) Electric Motor Mass (0.047 kg) Electric Motor Power (155.2 W) Electric Motor Mass (0.047 kg) Electric Motor Power (155.2 W) Electric Motor Mass (0.047 kg)	Battery Mass (4.171 kg) Battery Mass (4.173 kg) Battery Mass (4.179 kg)	—
Schomann [7]	Torque (RC/MP) (1.09/1.51) Engine Mass (0.77 kg)	Torque (EF/MP) 0.43/0.66 Electric Motor Mass (0.61 kg)	Battery Mass (3.89 kg)	Propeller Mass (0.18 kg)
Rippl [8]	Engine Mass (62.25 kg)	Electric Motor Mass (35.86 kg)	Battery Mass (84 kg)	—
Harmon [9]	Engine Power (837 W) Engine Mass (1.13 kg)	Electric Motor Power (114 W) Electric Motor Mass (0.16 kg)	Battery Mass 2.2 kg	Propeller Mass (0.17 kg)

Table 3.
Sizing comparison result of the hybrid propulsion characteristics of the HE – UAV achieved by various authors.

the authors used this control strategy to optimize the fuel consumption of HE-UAVs. Simulations show that a HE – UAV could achieve a fuel-saving potential of 6.9, 7 and 6.5%, respectively, compared to an Internal combustion Engine (ICE) -powered aircraft when using this control strategy.

Another controlled strategy used by researchers is the Fuzzy Logic Supervisory controlled strategy which is used as an energy management strategy to minimize the fuel consumption of HE-UAVs. Lei et al. [12], Xie et al. [13], and Bai et al. [14] employed a Fuzzy Logic supervisory controller to minimize the energy usage of a HE – UAV during its mission. Through Simulation, the authors showed that a HE – UAV could potentially achieve fuel savings of 5.9, 11, and 22%, respectively, compared to an ICE-powered UAV.

1.3 Hybrid: Electric UAV testbeds

After the sizing and modeled simulation of the UAV/Aircraft a final aspect in the construction of a HE-UAV is to build a testbed that seamlessly integrate the hybrid powertrain components of the HE – UAV. Various researchers has attempted to build such tested from the literature [15–25]. These were developed to validate the results obtained from the sizing of their HE-UAVs, and the results of numerical simulation models developed through a control strategy (energy management strategy) to optimize HE – UAVs performance for given mission profiles. Aksland [15] created a hybrid powertrain tested to perform a validation study of his mathematical simulation models and control algorithm which he developed for a HE – UAV. The testbed consists of an Energy Storage System (BMS, Battery Pack), Genset (Engine Starter/Generator, Hybrid ESC), Braking System (Power Supply, LP Battery Pack, Brake Motor, Brake ESC, Filter Box and Load Bank), Drivetrain System (Hybrid ESC, Propeller Motor) and DCDC system (DCDC ESC, Filter Box, Electronic Load). Subs section of the testbed was constructed for each of the components to validate each of the components. The model predicted a good validation with experimental data from the testbed. **Figure 1** shows the testbed constructed by the author.

Another study was done by Savvaris et al. [16], where he constructed a hybrid power testbed to test the control algorithms and mathematical simulation models, he developed for a HEPS for a light aircraft. The testbed consisted of an engine, motor/generator, fuel tank, and auxiliary component. A good performance match was obtained between the simulation models developed and experimental results obtained from the hybrid power tested. **Figure 2** shows the constructed hybrid power test bed.

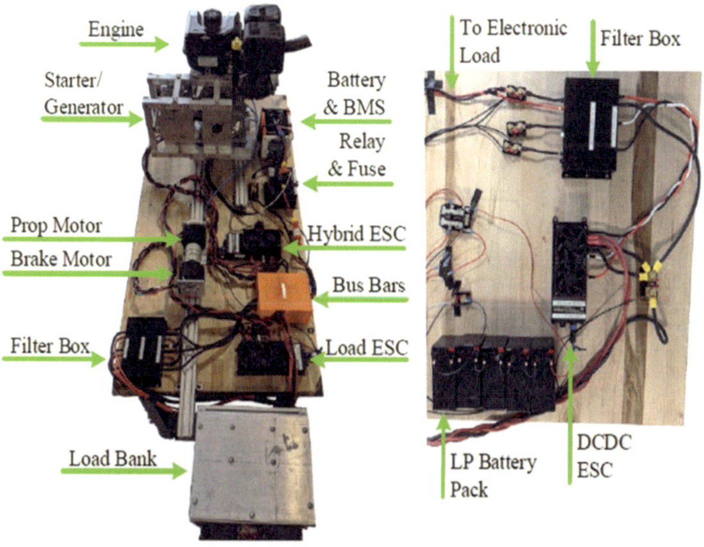

Figure 1.
Detailed labelling of HE – UAV powertrain testbed component [15].

Figure 2.
Completed testbed with the engine, motor/generator, fuel tank and all the auxiliary components installed [16].

Glassock et al. [17] constructed a prototype aircraft of a hybrid powerplant to show the feasibility of a hybrid powerplant for a small Unmanned Aerial System (UAS). The author began by developing a sizing methodology for selecting the components for building the hybrid power plant. The hybrid power plant was constructed and tested from the sizing results using a small UAS internal combustion engine (10cc methanol two-stroke) and a 600 W brushless direct current (BLDC) motor. Through his experiments, the author showed that a hybrid powerplant could improve the overall mission effectiveness and propulsive efficiency of a small UAS. **Figure 3** shows the constructed testbed.

Ausserer et al. [18] integrated a hybrid propulsion system on a small remotely piloted aircraft to evaluate its effectiveness for extended flight durations. The author started by constructing a testbed hybrid power plant (consisting of the Honda GX25 IC engine, starter motor, Maxon DC Motor and battery) of the propulsion system by performing a dynamometer test and wind tunnel test. The aircraft was retrofitted with this new propulsion system to evaluate its effectiveness for extended flight duration. A similar study was done by Dehesa et al. [19] where he developed a serial hybrid power test bed to validate his control algorithm on a series – HE – UAV for given mission profiles. The constructed testbed consisted of (a starter motor, generator, a 3 W-28i engine, a motor, battery, battery management system and propeller arranged in a series configuration).

Molesworth [20] developed a hybrid–electric remotely piloted aircraft testbed for validating the hybrid–electric system and control procedures he developed for a small remotely piloted aircraft. The testbed was built using the commercially off-the-shelf components of the ICE, Electric Motor, battery and propeller. The testbed was used to verify the functionality of the RPA hybrid powertrain in ICE Mode, EM Mode, the transition from ICE Mode to EM Mode, EM Mode to ICE Mode and dual Mode.

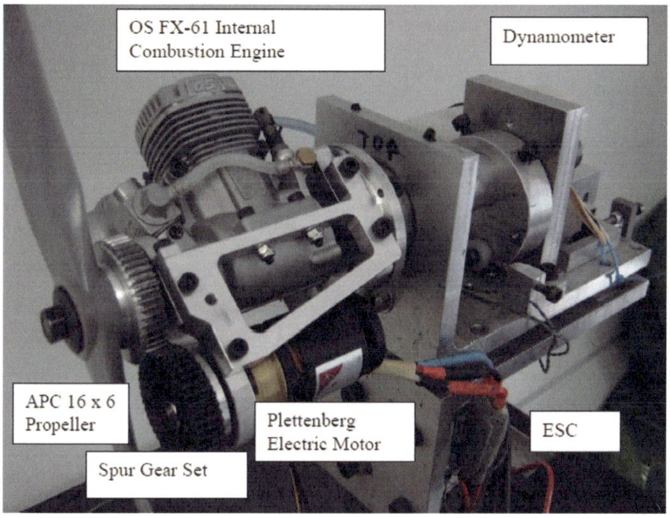

Figure 3.
Detailed labelling of a hybrid power testbed [16].

All modes functioned correctly, as predicted from Simulations of the hybrid–electric powertrain and control methodology.

Koster et al. [21] developed a testbed to validate a model of a hybrid powertrain configuration by integrating a hybrid powertrain propulsion system into the fuselage of the airframe of a HELIOS aircraft designed by the university. Modeling the hybrid powertrain configuration and constructing the hybrid powertrain testbed (Gearbox, ICE, EM and battery) was done. A helios aircraft was used as reference aircraft to retrofit the hybrid powertrain, and a flight test was performed. Another study was developed by Koster et al. [22] designed a testbed to validate the design of a UAS with Hybrid propulsion system (Hyperion aircraft). The testbed and aircraft were developed to validate the simulation results obtained using controllers. The authors did not display any conclusive result in the article.

Matlock et al. [23] developed a modular testbench of a hybrid propulsion system to compare the theoretical results obtained from the simulation of the hybrid framework of a hybrid propulsion system which is used to simulate various mission profiles of an aircraft developed in MATLAB with experimental results. The components of the hybrid propulsion system were individually modeled (Propeller, lithium polymer batteries, electric motors and internal combustion engine). A hybrid power testbed was constructed, and experiments were done to validate the results obtained from the hybrid framework. A similar study was done by Boggero et al. [24] developed a virtual testbench model, a parallel hybrid electric propulsion system for UAV to select its components and construct a real test bench of the hybrid UAV. The author started by creating models of the sub-system of the hybrid propulsion system in a Matlab Simulink environment. i.e. Electric Motor, Propeller, Battery ICE. Simulations were then performed by considering a given mission profile. Another study was done by Sliwinski et al. [25] developed a methodology for retrofitting a hybrid electric propulsion system (HEPS) into an Unmanned Aerial Vehicle. The Aerosonde UAV was considered reference aircraft to retrofit the hybrid propulsion

Author	IC engine	Electric motor	Converter	Propeller	Battery
Aksland [15]	19N1 Series Engine	Propulsion Motor Neu 8038-105(Test Motor) Dynamometer Motor Neu 8038-105 (Drive Motor) ZeroMax SC055R (Coupling) Hybrid ESC (Speed Contoller)	DCDC ESC and Filter Box	—	16S7P Battery Pack Samsung 18,650 Cells (Battery) TI Evaluation Board Bq76PL455-Q1 (BMS)
Glassock et al. [17]	OS FX-61 Internal Combustion Engine	Plettenberg Electric Motor	DC Generation	APC 16 x6 Propeller	Flight Power EV028
Ausserer [18]	Honda Motor Company	AXI Model Motors	R/C Servo to Analog Converter	APC Propellers	Thunder Power RC
Dehesa et al. [19]	3W-28i Single cylinder, 2-stroke engine	KDE 7215XF-135 brushless DC motor	AC power into DC power	APC 20x10 propeller	Turnigy Graphene Panther 6 s 22.2 V, 5 Ah LiPo battery
Molesworth [20]	Honda GX35 Engine	AXI Electric Motor	—	2 – bladed 18x12 APC propeller	Lithium Polymer Batteries
Koster et al. [21]	O.S. 0.46LA Engine Traxxas TRX 3.3 Engine	AXi 4120/20 Electric Motor	—	—	Two Max Amps 22.2 V 10 Amp Li-Po batteries
Matlock et al. [23]	4-stroke Saito FG36 gas engine	TG12-4 thermoelectric generator	—	—	LiPo batteries
Boggero et al. [24]	8-cylinder Diesel engine	EM2 Electric Motor	—	—	—
Sliwinski et al. [25]	Enya R120 4C Piston Engine	Plettenberg HP320/30 Electric Motor	—	—	Thunder Power Li-Po Air Battery

Table 4.
Comparison of the specification of testbed components for the construction of the HE UAV testbed from different authors from literature.

system. Its specifications were used to determine the flight dynamics and power requirement characteristics necessary to estimate the range and endurance of the retrofitted aircraft. **Table 4** shows a comparison of the component's specification used for the construction of the HE – UAV testbed selected from literature, **Table 5** shows a comparison of the different drone specification for the construction of the HE – UAV testbed and **Table 6** shows a comparison of testbed equipment for the construction of the HE – UAV testbed.

Author(s)	Drone size	Powerplant Engine	Cruise speed	Wingspan
Savvaris et al. [16]	65 kg Aegis UAV Berkut 360 Canard Wing Light Aircraft	205 hp. (Maximum Engine powerplant)	50knots	26′8″
Ausserer [18]	RPA build by AFIT	Honda GX35 engine	—	3.66 m
Dehesa et al. [19]	18 kg AAI Aerosonde UAV	—	—	2.89 m
Molesworth [20]	RPA build by AFIT	Honda GX35(35 cc) 4-stroke gasoline engine	—	12 ft
Koster et al. [21]	Aircraft Designed by Daniel Webster College team	O.S. 0.46LA Engine Traxxas TRX 3.3 Engine	—	13 ft
Koster et al. [22]	Hyperion Aircraft	Gas Electric Engine	27 m/s	3 m
Sliwinski et al. [25]	Aerosonde RPAS	Enya R120 4C Piston Engine	—	—

Table 5.
Comparison of reference drone specification for the construction of the HE – UAV testbed by various authors from literature.

Author(s)	Instruments for testbed construction	Function
Glassock et al. [17]	Reaction Type dynamometer Yokogawa DL50 Strain	Measures the Torque produced by the hybrid Powerplant
	DL50 Strain Guage Amplifier	Enables rebalance of the gauges at any time during the experiment
Ausserer [18]	DYNOmite Mini Eddy Dyno96V Dynamometer	Measures Torque, Speed and power produced By the hybrid powerplant
	Max Machinery 213 piston helical flow meter	Measures Engine Fuel Flow

Table 6.
Comparison of testbed equipment's for the construction of the HE – UAV testbed from literature.

2. Conclusion

This paper attempted to explore the feasibility of constructing a prototype of hybrid power testbed of a HE-UAV. A number of technical papers in the research area were obtained from literature were selected which were addressed in the relevant sub sections of the paper; component sizing, numerical simulations and hybrid power testbed construction. A comparison between each author result was done by comparing sizing results achieve by each author from a reference drone size. In addition, a comparison of the testbed component for constructing the HE – UAV and testbed equipment used by each author was done. The results from this research shows that a HE – UAV testbed can be achieve provided stringent measures are done in determining the size of the component for a given drone size, and a careful selection of the components from the sizing result for the testbed construction.

Acronyms and abbreviations

HE	UAV hybrid electric unmanned aerial vehicle
UAV	unmanned aerial vehicle
UAS	unmanned aerial system
HEPS	hybrid electric propulsion system
ICE	internal combustion engine
EM	electric motor

Author details

Lionel Fouellefack
Department of Mechanical and Aeronautical Engineering, University of Pretoria, Pretoria, South Africa

*Address all correspondence to: u18128361@tuks.co.za

References

[1] Fouellefack L, Smith L, Kruger M. Development of a novel supervisory controller on a parallel-hybrid powertrain for small unmanned aerial system. Aircraft Engineering and Aerospace Technology. 2023;**95**:401-415

[2] Xie Y, Savvaris A, Tsourdos A. Sizing of hybrid electric propulsion system for retrofitting a mid-scale aircraft using non-dominated sorting genetic algorithm. Aerospace Science and Technology. 2018;**82-83**:323-333

[3] Donateo T, Ficarella A, Spedicato L. A method to analyse and optimize hybrid electric architectures applied to unmanned aerial vehicles. Aircraft Engineering and Aerospace Technology. 2018;**5**:828-842

[4] Donateo T, Ficarella A. Designing a hybrid electric powertrain for a unmanned aircraft with a commercial optimization software. SAE International Journal of Aerospace. 2017;**10**:1-11

[5] Pornet C, Patrick V, Seitz A, Schmitz O. Methodology for sizing and performance assessment of hybrid energy aircraft. Journal of Aircraft. 2013;**52**:341-352

[6] Hiserote M. Analysis of hybrid-electric propulsion system designs for small unmanned aircraft systems. In: 8th Annual International Energy Conversion Engineering Conference, Ohio, USA; pp. 66-87

[7] Schomann J. Hybrid-electric propulsion systems for small unmanned aircraft. [Doctoral Thesis]. Munich, Germany: Technical University of Munich; 2014

[8] Rippl D. Sizing analysis for aircraft utilizing hybrid – Electric propulsion system. [Master's Thesis]. Ohio, USA: Air Force Institute of Technology; 2011

[9] Harmon G. Neural network control of a parallel hybrid – Electric propulsion system for a small unmanned aerial vehicle. Neural Networks. 2005;**18**:772-780

[10] Xie Y, Savvaris A, Tsourdos A, Laycock J, Farmer A. Modelling and Control of a Hybrid Electric Propulsion System for Unmanned Aerial Vehicles. Institute of Electrical and Electronics Engineer (IEEE Aerospace Conference); 2018. pp. 1-13

[11] Hung J, Gonzalez F. On parallel hybrid-electric propulsion system for unmanned aerial vehicle. Progress in Aerospace Sciences. 2012;**51**:1-17

[12] Lei T, Wang Y, Jin X, Min Z, Zhang X, Zhang X. An optimal fuzzy logic based energy management strategy for a fuel cell/battery hybrid power unmanned aerial vehicle. Aerospace. 2022;**2**:115

[13] Xie Y, Savvaris A, Tsourdos A. Fuzzy logic based equivalent consumption optimization of a hybrid electric propulsion system for unmanned aerial vehicles. Aerospace Science and Technology. 2019;**85**:13-23

[14] Bai M, Yang W, Song D, Kosuda M, Kelemen M. Equivalent consumption minimization strategy based on fuzzy logic control for the energy management of hybrid unmanned aerial vehicle. In: Proceedings of the international Conference on Applied Energy, Kosice, Slovakia; 2021

[15] Aksland C. Modular modelling and control of a hybrid unmanned aerial Vehicle's powertrain. [Master Thesis].

Urbana, Illinois: University of Illinois at Urbana-Champaign; 2019

[16] Savvaris A, Xie Y, Wang L, Wang S, Tsourdos A. Control and optimization of hybrid – Electric propulsion system for light aircraft. The Journal of Engineering. 2018;**13**:478-483

[17] Glassock R, Hung J, Gonzalez L, Walker A. Design, modelling and measurement of hybrid Powerplant for unmanned aerial systems. Australian Journal of Mechanical Engineering. 2008;**2**:69-78

[18] Ausserer J. Integration, testing, and validation of a small hybrid-electric remotely – piloted aircraft. [Master's Thesis]. Ohio, USA: Airforce Institute of Technology; 2012

[19] Dehesa D Jr. Study of control schemes for series – Hybrid-electric powertrain for unmanned aerial systems. The Journal of Engineering. 2020;**13**:478-483

[20] Molesworth M. Rapid prototype development of a remotely – Piloted aircraft powered by a hybrid – Electric propulsion system. Master's Thesis. Ohio, USA: Airforce Institute of Technology; 2012

[21] Koster J, Humbargar C, Serani E, Velazco A, Hillery D, Makepeace L. Hybrid electric integrated optimized system (HELIOS) design of a hybrid propulsion system for aircraft. In: 49th AIAA Aerospace Sciences Meeting Including the New Horizons Forum and Aerospace Exposition, Orlando, Florida; 2011. pp. 1011

[22] Koster J, Balaban S, Hillery D, Humbargar C, Nasso D, Serani E, et al. Design of a blended wing body UAS with hybrid propulsion. In: ASME International Mechanical Engineering Congress and Exposition, Orlando, Florida. Vol. 54877. 2011. pp. 331-337

[23] Matlock J, Warwick S, Sharikov P, Richards J, Suleman A. Evaluation of energy efficient propulsion technologies for unmanned aerial vehicles. Transactions of the Canadian Society for Mechanical Engineering. 2019;**43**:481-489

[24] Boggero L, Corpino S, De Martin A, Evangelista G, Fioriti M, Sorli M. A virtual test bench of a parallel hybrid propulsion system for UAVs. Aerospace. 2019;**7**:77

[25] Sliwinski J, Gardi A, Marino M, Sabatini R. Hybrid – Electric propulsion integration in unmanned aircraft. Energy. 2017;**140**:1407-1416

Chapter 7

Integrating Drones across the Curriculum at Stephen F. Austin State University

David L. Kulhavy, Daniel R. Unger, I-Kuai Hung, Victoria M. Williams, Yanli Zhang and Reid Viegut

Abstract

Faculty within the Arthur Temple College of Forestry and Agriculture at Stephen F. Austin State University are heavily involved with using drones to quantify and qualify forests and natural resources. Drones are involved across the entire curriculum within the College of Forestry and Agriculture to enhance the educational experience of all students. Emphasis is placed on integrating drones, and their effective use in forest and natural resource management endeavors, within all three main focus areas of teaching, research, and service. In addition to the traditional three education focus areas, faculty within the College of Forestry and Agriculture also help train future drone pilots as part of their educational experience.

Keywords: drone, forest, natural resource management, education, drone pilot

1. Introduction

The use of drones is expanding in use within the natural resources community as a means to provide complex measurements of natural resources for teaching, research, and service activities. To produce foresters ready to be productive within society, they need to be effective at quantifying natural resources and have the ability to communicate this information clearly.

Natural resource education within the twenty first century, and the foreseeable future, must be relevant to the goals of society [1]. For natural resource managers, drones provide a useful tool when integrated into geospatial technologies to measure and predict forest and natural resources at the stand, forest, and landscape levels. Increasing a student's skill level and educational knowledge while integrating math and statistics to analysis real-world natural resource-based situations is paramount in today's world. Bridging the gap between a dedication to lifelong learning with necessary hands-on skills is of high importance.

Drones are relatively new to the process of acquiring data within the natural resources' profession, and increased emphasis needs to focus on training students with this new technology. Upon graduation students need to be ready to engage with society and have the capacity to deal effectively with the complex economic, political, ecological and social issues natural resource managers deal with every day. Natural resource graduates must be prepared to enhance the health of the natural resource environment through sustainable management, conservation, and protection of earth's resources. As technology and drones continue to change and evolve, educators need to adjust and adapt their teaching methods to meet these new societal challenges [2, 3].

Within the Arthur Temple College of Forestry and Agriculture (ATCOFA) at Stephen F. Austin State University (SFASU) there are four faculties dedicated to the integration of drone technology across the entire curriculum. The four faculties are collectively called the "Drone Squad" as they focus their teaching, research, and service activities on using drones to effectively quantify and qualify forests and natural resources. The goals of the faculty are to meet the college's mission objectives of making a difference, work outdoors, while using high-end technology.

Drones, and the imagery they collect, are increasing in popularity and use and are the future of remotely sensed data acquisition. Drones allow you to acquire the imagery and videos of what you want and when you want it collected. In addition, one of the most appealing aspects of using a drone to acquire remotely sensed data is that the user controls the spatial, spectral, radiometric, and temporal resolution of any data collected for forest composition, structure, volume and growth [4, 5].

Historically foresters and natural resource managers have relied on satellite-based platforms for imagery. Although effective at the landscape level, satellite-based platforms with their traditional poor spatial resolution have been ineffective in providing high spatial resolution data for visual interpretation at the individual tree level. Drones with their high spatial resolution controlled by the user can identify individual trees and plant species. In addition to providing a traditional nadir or 90-degree angle perspective, drones can be flown interactively and with a movable camera that can also provide the user with off-nadir or side perspectives of earth surface objects such as the side of a tree which can be very useful in tree stem assessments.

The objective of this chapter is to present the reader how drones can be integrated across a natural resource-based curriculum to increase a student's skill set. Examples of how drones are integrated across the curriculum at SFASU are presented within the context of the three main focus areas of an academic institution being teaching, research, and service endeavors. Finally, a fourth focus area demonstrates that SFASU faculty are heavily involved with training current students to be Federal Aviation Administration (FAA) certified drone pilots.

2. Teaching–Integration within ATCOFA classes

Drones are integrated across the entire curriculum within ATCOFA with a focus on forests and natural resource management activities [4]. Specific examples are given that highlight how effective drones can be to quantify and qualify natural resources, which gives the reader a perspective on how broad the applications can be in the real world.

For example, drones can be used to fly a recreational hiking trail to visually highlight the trail for individuals not able to physically walk the trail themselves or to document the need for trail maintenance in the future (**Figure 1**, Video 1, https://bit.ly/3tYlx3U).

In **Figure 2**, drones are used to capture and record the felling of a tree on the SFASU campus as an aid to help forestry students understand the complex nature

Figure 1.
Example of a drone flying and recording a hiking trail in east Texas.

Figure 2.
Drone capturing the felling of a diseased tree limb.

of felling a diseased tree within an urban forestry setting (Video 2, https://bit.ly/3tYlx3U). In this particular example, a drone was used to record the felling of a limb as part of a classroom exercise.

Part of the goal of ACTOFA is to provide interactive outside demonstrations for students visiting the SFASU campus to get them excited about college and natural resource management as a future possibility. **Figure 3** demonstrates the ATCOFA college dean talking with prospective students about the benefits of a natural resource-based education. This drone video was taken for future recruitment of undergraduate students (Video 3, https://bit.ly/3tYlx3U).

An important component of curricula within ATCOFA is the wildlife major where students study wildlife science. Drones can be an effective tool in helping to quantify

Figure 3.
Drone video of an interactive outside natural resource demonstration.

Figure 4.
Counting turtles in LaNana Creek, Nacogdoches, Texas.

and identify wildlife species from a distance that might not be approachable in person. **Figure 4** demonstrated the use of drone imagery to quantify turtles within a local creek in Nacogdoches, Texas (Video 4, https://bit.ly/3tYlx3U).

An additional component of the education received by the students within ATCOFA, and highly praised by the faculty and future employers, is the ability of the faculty to introduce students to ecosystems and experiences they typically do not receive on campus. To that end, faculty travel the country and record drone imagery of different locations to enhance a student's educational experience. In **Figure 5**, faculty recorded and embedded in course work an example of a beaver dam that was destroyed by early spring snow runoff in northern Utah (Video 5, https://bit.ly/3tYlx3U).

For foresters, especially in their undergraduate education, it is important to not only understand trees but the conditions of the forest floor. **Figure 6**, a video of Kamiak Butte in eastern Washington state north of Pullman, Washington, shows the

Figure 5.
Beaver dam destroyed by snow run off in northern Utah.

Figure 6.
Forest floor evaluation on Kamiak Butte, Washington.

influence of the forest floor in an arid intermountain ecosystem (Video 6, https://bit.ly/3tYlx3U).

In addition to wildlife majors within ATCOFA, the faculty also teach a number of outdoor recreation students involved with natural resource management activities. **Figure 7** was collected by faculty to demonstrate the conditions of recreational sites across the United States to demonstrate not only how they vary from location to location, but also their similarity (Video 7, https://bit.ly/3tYlx3U). **Figure 7** demonstrates a recreation site within Scenic Beach State Park in western Washington about an hour west of Seattle.

For the traditional foresters with ATCOFA, it is paramount that they not only understand trees and forest ecosystems but that they also understand open range

Figure 7.
Recreation site within Scenic Beach State Park, Washington.

Figure 8.
Range assessment in northern Utah.

ecosystems typical of the western United States. **Figure 8** was collected to demonstrate open range conditions and the sparseness of vegetation in the arid western United States that students cannot be introduced to on the campus of SFASU. **Figure 8** demonstrates a video that displays the vegetation distribution of open range in the Unita-Wasatch-Cache National Forest in northern Utah (Video 8, https://bit.ly/3tYlx3U).

Figure 9 demonstrates a video obtained of the Hell Roaring Creek fire in central Idaho within the Sawtooth National Recreation Area. This particular imagery was flown to demonstrate to the ATCOFA forestry students the damaging effects of fire within a Lodgepole Pine ecosystem (Video 9, https://bit.ly/3tYlx3U).

Figure 10, also flown for our forestry undergraduate students, demonstrates the uniqueness of Aspen trees relative to other conifer trees within a forest ecosystem.

Figure 9.
Fire damage within a Lodgepole Pine ecosystem.

Figure 10.
Aspen trees with a forest landscape in northern Utah.

Aspen trees grow in clumps relative to other trees within an ecosystem, which can be readily seen from a spatial perspective giving students a unique perspective when trying to identify individual tree species with a forested landscape (Video 10, https://bit.ly/3tYlx3U).

While introducing our outdoor recreation majors to different recreation sites across the United States, it is important to capture imagery from different ecosystems. **Figure 11**, in contrast to **Figure 7**, depicts a recreation site within the Sawtooth National Recreation Area in central Idaho. This area represents an arid intermountain location different from the recreation area depicted in **Figure 7**. This image represents a recreation site on the shore of Pettit Lake south of Stanley, Idaho in the Sawtooth National Forest (Video 11, https://bit.ly/3tYlx3U).

Figure 11.
Recreation site on the shore of Pettit Lake, Sawtooth National Forest.

Another important component of a forestry education in the United States is for students to understand the effect of beetle damage on a forest. **Figure 12** depicts the effect of mountain pine beetle, *Dendroctonus ponderosae*, damage on Lodgepole Pine in the proximity to Little Redfish Lake in the Sawtooth National Forest in central Idaho (Video 12, https://bit.ly/3tYlx3U).

Figure 13 shows our undergraduate students a recreation site on the Snake River south of Jackson, Wyoming. The video shows a popular recreation activity in the western United States, river rafting, which our students cannot experience in east Texas (Video 13, https://bit.ly/3tYlx3U).

Our recreation students are also interested in historical artifacts and buildings scattered throughout our forests (**Figure 14**). Not only in their preservation, but in

Figure 12.
Lodgepole pine beetle damage, Sawtooth National Forest.

Figure 13.
River rafting on the Snake River south of Jackson, Wyoming.

Figure 14.
Historic cabin deep within the Stanley Basin in central Idaho.

their education as a means to help others understand the importance of historical artifacts. This video was flown to introduce our students to typical historical dwellings found within forests of the United States, in this particular case an historical cabin, that exists within the Stanley Basin in central Idaho (Video 14, https://bit.ly/3tYlx3U).

Figure 15 is an important video that was flown to introduce our spatial science students to the effects of light interaction on the landscape and its effect on vegetation distributions. This is a drone video of Spawn Creek within Logan Canyon of northern Utah. It was flown to demonstrate the effect of intense solar radiation by drying out one side of a southerly exposed hill side versus a more shaded hill side (Video 15, https://bit.ly/3tYlx3U), emphasizing the importance of aspect in plant growth.

Figure 15.
Effects of solar radiation on vegetation distributions.

Figure 16.
Forest floor post fire assessment.

Our last example of a drone video flown for teaching purposes depicts an image of the Sawtooth National Recreation Area (**Figure 16**). This was acquired to introduce our students to the effects of fire from a nadir or birds-eye perspective for the ease of quantifying downed timber as a result of a recent forest fire (Video 16, https://bit.ly/3tYlx3U).

3. Research–Quantifying natural resources

In addition to integrating drones into the curriculum, faculty also are heavily involved in drone related research projects. Research has involved many facets of drones and their uses, with an emphasis on how to best use drones in quantifying forests and natural resources.

DOI: http://dx.doi.org/10.5772/intechopen.1002705

Figure 17.
Example of 3D landscape modeling.

Figure 17 shows the results of a research project utilizing drone imagery to produce a 3D model of the landscape [6]. The results of 3D modeling can be used to visually portray a landscape to aid in the visual assessment of natural resources.

Another factor our forestry students are concerned about is insect damage on a forest. Many research projects within ATCOFA have involved using the high spatial resolution of drone acquired imagery to assess insect damage at a level of detail not attainable by traditional satellite-based platforms. **Figure 18** is an example of using drone imagery undertaken by faculty within ATCOFA to assess the impact of forest damage due to the *Ips* bark beetles.

In addition to using drone imagery to create 3D models of the landscape, drones can be used to research their effectiveness at creating 3D models of human created earth surface features and their impact on the natural resource ecosystem. **Figure 19** portrays ATCOFA research looking into the effects of human created surface features and building on the surrounding environment from a visual perspective.

Figure 20 portrays current research on the campus of SFASU involving the spatial distribution of litter on campus. High spatial resolution drone imagery was flown to

Figure 18.
Ips bark beetle forest damage assessment.

Figure 19.
Using 3D terrain modeling to assess a natural resource environment.

not only quantify litter on the SFASU campus, but to model the spatial distribution as an aid to increase awareness of the litter problem on campus in an effort to minimize its negative impact on the environment and to plan litter clean up.

One of the most important aspects of using drones is to combine individual drone images together to create an orthophoto mosaic of an area. By acquiring drone images over a study area with sufficient overlap and side lap, the individual drone images can be stitched together to create a highly accurate composite image or orthophoto mosaic. Considerable research has been undertaken by faculty within ATCOFA to assess the

Figure 20.
Hazard rating litter on the campus of SFASU.

accuracy of drone created orthophoto mosaics and have found that drone created orthophoto mosaics are highly accurate and can be used as base maps in geospatial science related products with a high degree of accuracy and confidence (**Figure 21**) [7, 8].

Figure 22 depicts a research project assessing the utility of using drone imagery to hazard rate urban trees for disease and infestations. Traditionally this has been obtained from the ground using a visual assessment, but drone imagery allows the user to acquire imagery closer than obtainable from the ground and from angles not previously attained. ATCOFA research has shown that drones can provide accurate tree hazard rating equal to or better than traditional field assessments and in a timely and efficient manner (**Figure 22**) without significant differences [9].

The last research project presented in this chapter, which is portrayed in **Figure 23**, shows the use of change detection methodology employed to identify landscape change overtime. This particular project used current drone imagery compared to an historical creek in Nacogdoches, Texas digitized from 1939 historic aerial photographs to identify the location of current landscape features relative to the location of the historic creek bed locations. This drone research project shows how current high-end technology can be merged with historical data to provide new insights and implications for current natural resource management perspectives.

Figure 21.
Orthophoto mosaic created of the campus of SFASU.

Figure 22.
Assessing tree hazard with a drone.

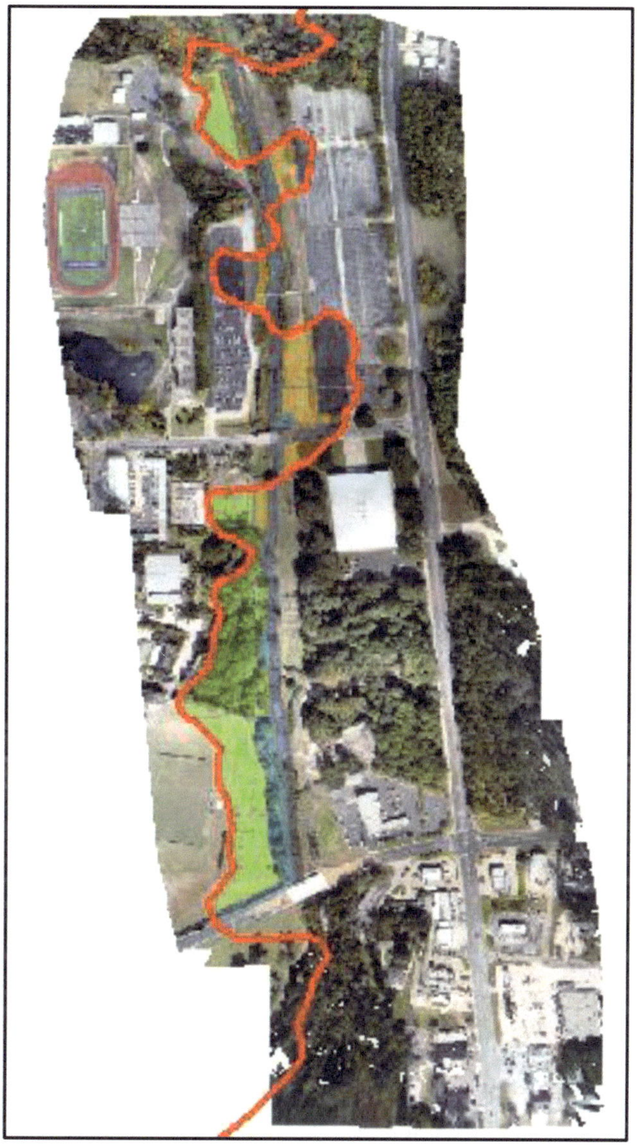

Figure 23.
Drone imagery merged with an historic creek river channel.

4. Service–Local, state, and the world

In additional to traditional teach and research activities, faculty within ATCOFA also integrated drone technology into service project benefiting local communities and society in general [6, 10]. Faculty in ATCOFA have found the use of drones to be a useful took in aiding local communities acquire useful and pertinent information.

Figure 24 depicts a local service project where the faculty flew a drone to create an orthophoto mosaic for a local park in Lufkin, Texas. The drone was flown to acquire

Figure 24.
Orthophoto mosaic of a city park in Lufkin, Texas.

individual drone images that were then utilized to create a highly accurate orthophoto mosaic for the city park. The city of Lufkin, Texas will use the created orthophoto mosaic in future planning activities for proposed expansion of an existing zoo adjacent to the city park [8, 11].

Figure 25.
Hall 16 demolition on the campus of SFASU.

Figure 26.
ROTC field exercise–hidden soldiers at base of trees.

The campus of SFASU has recently begun the demolition and construction of various buildings on campus. As an aid to the campus community, faculty and graduate students within ATCOFA have begun to fly the demolition and construction of various buildings on campus. Goals are to not only to provide visual documentation of the demolitions and reconstruction, but to demonstrate to students on campus of the utility of using drones to document urban change (**Figure 25**).

A recent service project undertaken by the faculty within ATCOFA was to partner with the ROTC program on the SFASU campus to demonstrate the utility of drone acquired imagery for use with tactical training of Army officers within the ROTC program. Faculty traveled with the ROTC staff to a recent training site and aided their training by providing real time imagery to officers in training to demonstrate the utility of real-time battle field imagery. Faculty were paired with individual ROTC cadets with a drone to provide them with real-time imagery of a battlefield to aid their preparation and battlefield assessments (**Figure 26**).

5. Training future drone pilots

The fourth focus area faculty are heavily involved with at SFASU involves training students in not only how to use drones to effectively quantify and qualify forests and natural resources, but to train individual students to be FAA certified drone pilots [4, 12]. Thereby, increasing their skill level and employment prospects upon graduation.

ATCOFA faculty closely follow FAA guidelines while training student drone pilots. Faculty merge the process of physically flying drones with traditional classroom education on drone pilot rules and regulations. Our university recently added a course on drones and Geographic Information Systems (GIS) and the use of drones continues to be integrated across the curriculum. The faculty's intent was to introduce students to the use of high-end technology within the natural-resource-based decision-making process [4, 12].

A drone's unique ability to acquire remotely sensed aerial imagery and videos of natural resources at the time and place of the user's control is one of the main advantages of using a drone in the field. The user, rather than relying on traditional satellite-based imagery which has been the mainstay of remote sensing since 1972, can now control the spatial, spectral, radiometric and temporal resolution of data needed for a particular natural resource.

By integrating multiple different digital images obtained with a drone, the user can now create their own orthophoto mosaic of a geographic area in a few hours while controlling the spatial environment of the mosaic product. To create a mosaic project, the drone is unpacked and assembled in a matter of minutes, then the project is defined with a user-controlled application following which the drone imagery are obtained with an app. The images are eventually downloaded using drone-specific software to create an orthophoto mosaic. For a geographic area encompassing 30 ha, the process can take less than 3 hours from start to completion [4].

Before students can fly a drone, a four-step process was developed as part of a Mentored Undergraduate Scholarship program at Stephen F. Austin State University [4]. Successful operation of a drone was based on a four-part process that included three intermediate steps culminating in a capstone focus at the end. The three intermediate steps represented first learning how to assemble a drone, second involved learning how to complete a controlled drone flight, and the third step involved learning how to use a drone to obtain three multiple points of measurements from both imagery and video. The culminating capstone at the end involved learning how to synthesize drone information and knowledge to properly quantify and qualify natural resources ([4], p. 403).

For intermediate step 1, the operator has to assemble the drone ready for flying, then pass a safety inspection by the instructor. For intermediate step 2, the pilot needed to demonstrate how to pilot a drone and successfully record an image and video. For intermediate step 3, the drone operator needed to download images and videos into geospatial science software to effectively quantify natural resources. For the capstone assessment, the pilots needed to demonstrate the application and user of drone acquired imagery and video into specific natural resource assignments. As a final capstone assessment, drone operators demonstrating proficiency in flying the drones were allowed to complete the FAA drone pilot examination to become a certified drone pilot [4].

With the ever-increasing use of more user-friendly drones, more compact size and lower prices, drone use in natural resource management will only continue to exponentially increase. Early student exposure and instructor led hands-on use of drones will develop the skill sets defined by Bullard et al. requiring the integration of technology with a traditional education [2, 3]. Drones allow this integration by permitting the user to acquire accurate information on forest composition, structure, volume and growth [6]. To meet these society goals, emphasis within ATCOFA is being placed on expansion of drones across the entire curriculum [4, 6, 12].

With increased emphasis on hands-on use of drones within ATCOFA, increased focus is place on service-learning activities [13]. Focus areas include being attentive to the community partner's mission and vision; understanding the human dimension of a community's work focus; being aware of a communities limited resources; accepting and sharing inefficiencies that limit productivity; considering the impact of lifelong agreements; and most importantly viewing the need to continue of progress as important [14].

These six focus areas were integrated into teaching, research and service with drones in the development of service-learning activities for the Port Jefferson History Center and Nature Center on use of drones in documenting the natural resource aspects of the site [6, 15]. In addition, a three-dimensional model of the Caddo House destroyed by a tornado was recreated as a remembrance of the event for the Caddo Nation celebration (**Figure 27**) [16]. Hands-on activities which included the estimation of heights of trees, buildings and light poles using drones were constructed in multiple natural resource courses (**Figure 28**) [10, 12, 16]. The major finding was that the drone needed to land and be reset between individual drone flights to acquire more accurate measurements. Repeated drone measurements of the same area over time indicated positional accuracy was highest at the center of the orthophoto mosaic compared to edge effects with increased overlap (**Figure 29**) [7, 17]. Drone use has also been emphasized in citizen science communication incorporating natural resource management endeavors [18].

Figure 27.
3D model of Caddo House destroyed by a tornado.

Figure 28.
Using a drone to estimate height accuracy [10].

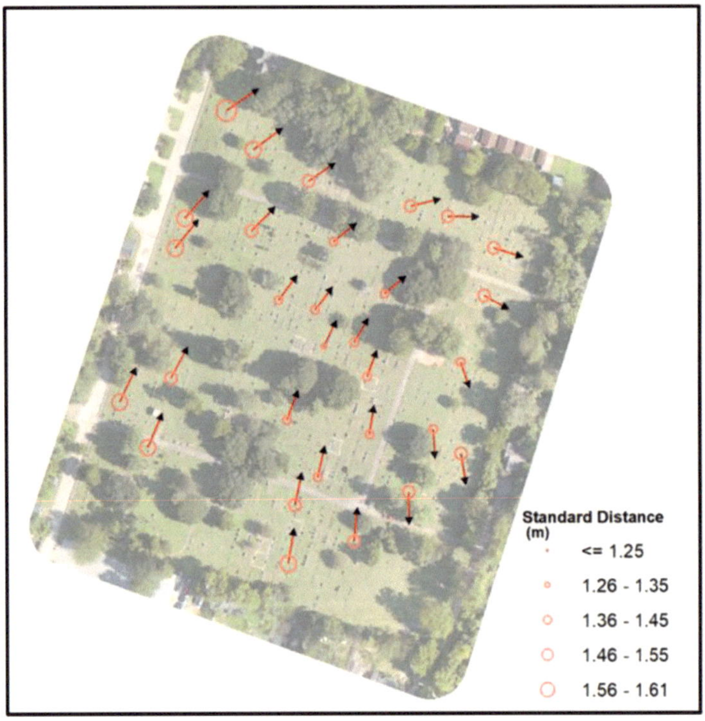

Figure 29.
Map of mean center locations with the magnitude of precision presented as the standard distance.

6. Conclusions

The integration of hands-on training in the use of drones emphasizes the mission of the Arthur Temple College of Forestry and Agriculture to produce "society ready" natural resource managers while maintaining excellence in teaching, research and service to enhance the health and vitality of the environment through sustainable management, conservation and protection of our forests and natural resources to promote the production and economic vitality of natural resources. The land ethic of work outdoors, make a difference and use high-end technology is enhanced through the use of drones as the students must use safe techniques and follow FAA regulations of UAS operation. The four-step process of drone assembly; controlled flight using a Pilot in Command and a Visual Observer; preparation of an orthomosaic using the proper applications and training; and the synthesis of drone information across landscapes, prepares students for safe and productive drone missions. To date, 30 students have taken and passed the FAA Unmanned Aerial System Pilot exam and taken the FAA TRUST test. The increase complexity and multitude of sensors of the drones means increased training and readiness for successful drone pilots.

Acknowledgements

The authors with to acknowledge the many students over the years who have participated in learning how to use drones to map, manage, and monitor forests and natural resources. It is through your valuable contribution and love of learning that has made the validation of introducing drones within the ATCOFA curriculum a success. This work was supported by McIntire Stennis funds administered by the Arthur Temple College of Forestry and Agriculture and a Center for Applied Research and Rural Innovation grant from Stephen F. Austin State University.

Conflict of interest

The authors declare no conflict of interest.

Author details

David L. Kulhavy[1*], Daniel R. Unger[1], I-Kuai Hung[1], Victoria M. Williams[1], Yanli Zhang[1] and Reid Viegut[2]

1 Arthur Temple College of Forestry and Agriculture, Stephen F. Austin State University, Nacogdoches, Texas, USA

2 Missouri Cooperative Fish and Wildlife Research Unit, University of Missouri, Columbia, Missouri, USA

*Address all correspondence to: dkulhavy@sfasu.edu

References

[1] Bullard SH. Forestry curricula for the 21st century—maintaining rigor communicating relevance, building relationships. Journal of Forestry. 2015;**113**:552-556. DOI: 10.5849/jof.15-021

[2] Bullard SH, Stephens Williams P, Coble T, Coble DW, Darville R, Rogers L. Producing "society-ready" foresters: A research-based process to revise the bachelor of science in forestry curriculum at Stephen F. Austin State University. Journal of Forestry. 2014;**112**:354-360. DOI: 10.5849/jof.13-098

[3] Bullard S, Coble D, Coble T, Darville R, Rogers L, Stephens Williams P. Producing "Society-Ready" Foresters: A Research-Based Process to Revise the Bachelor of Science in Forestry Curriculum at Stephen F. Austin State University [ATCOFA Monograph 1-2014]. Nacogdoches: Arthur Temple College of Forestry and Agriculture, Stephen F. Austin State University; 2014

[4] Unger DR, Kulhavy DL, Hung I-K, Zhang Y, Stephens WP. Integrating drones into a natural-resource curriculum at Stephen F. Austin State University. Journal of Forestry. 2019;**117**:398-405. DOI: 10.1093/jofore/fvz031

[5] Tang L, Shao G. Drone remote sensing for forestry and research and practices. Journal of Forest Research. 2019;**26**:791-797. DOI: 10.1007/s11676-015-0088-y

[6] Kulhavy DL, Unger DR, Grisham R, Coble D, Endsley G, Gannon M. Service learning for the Port Jefferson History and Nature Center: Senior capstone forestry course. Journal of Community Engagement & Higher Education. 2017;**9**:67-79

[7] Hung I-K, Unger D, Kulhavy D, Zhang Y. Positional precision analysis of orthomosaics derived from drone captured aerial imagery. Drones. 2019;**3**:Article 46. DOI: 10.3390/drones3020046

[8] Williams V, Unger DR, Kulhavy D, Hung I-K, Zhang Y. Comparing Drone2Map versus Pix4Dmapper when creating orthophoto mosaics over homogeneous land features. International Journal of Geospatial and Environmental Research. 2023;**10**:Article 2. Available from: https://dc.uwm.edu/ijger/vol10/iss1/2/

[9] Kulhavy D, Unger D, Hung I, Zhang Y. Comparison of AR.Drone quadcopter video and the visual CTLA method for urban tree hazard rating. Journal of Forestry. 2016;**114**:517-523. DOI: 10.5849/jof.15-005

[10] Unger DR, Hung I-K, Zhang Y, Kulhavy DL. Integrating drone technology with GPS data collection to enhance forestry students interactive hands-on field experiences. Higher Education Studies. 2018;**8**:49-62

[11] Unger DR, Hung I-K, Kulhavy DL, Zhang Y, Busch-Petersen K. Accuracy of unmanned aerial system (drone) height measurements. International Journal of Geospatial and Environmental Research. 2018;**5**:Article 6. Available from: https://dc.uwm.edu/ijger/vol5/iss1/6/

[12] Unger D, Kulhavy D, Busch-Petersen K, Hung I-K. Integrating faculty led service learning training to quantify height of natural resources from a spatial science perspective. International Journal of Higher Education. 2018;**5**:104-116

[13] Tinkler A, Tinkler B. Teaching across the community: Using service learning field experiences to develop culturally and linguistically responsive teachers. In: Jagla VM, Erickson JA, Tinkler AS, editors. Transforming Teacher Education Through Service Learning. Charlotte, NC: Information Age Publishing; 2013. pp. 99-117

[14] Latorre-Navarro E, Meier E. Strategies to fund community service capstone projects 2022 In: Capstone Design Conference; 6-8 June 2022; Dallas. 2022. p. 2. Available from: https://www.capstonedesigncommunity.org/sites/default/files/proceedings_papers/LatorreNavarroMeier_FundingCommunityServiceProjects_CDC22.pdf

[15] Zhang Y, Kulhavy D, Gerland J, Hung I-K, Unger D, Wen X, et al. Evaluating different UAS flight methods for 3D model generation and printing of a tornado destroyed cultural heritage: Caddo House in Texas. Drones and Autonomous Vehicles. 2023;**1**:10003. DOI: 10.35534/dav.2023.10003

[16] Kulhavy DL, Hung I-K, Unger DR, Viegut R, Zhang Y. Measuring building height using point cloud data derived from unmanned aerial system imagery in an undergraduate geospatial science course. Higher Education Studies. 2021;**11**:105-113

[17] Wiliams V, Unger D, Kulhavy D, Hung I-K, Zhang Y. Assessing drone mapping capabilities and increased cognitive retention using interactive hands-on natural resource instruction. Higher Education Studies. 2023;**13**(2):28-36

[18] Kulhavy D, Unger D, Hung I-K, Schalk C, Zhang Y, Viegut R. Integrating natural resources education through the use of unmanned aerial systems (drones). International Journal of Higher Education. 2022;**11**:143-152

Chapter 8

Leveraging Programmable Educational Drones, Robots and AI for Learning STEM, Computational Thinking and Higher Order Thinking in Schools in Rural Villages

John-Thones Amenyo and Wolanyo Kpo

Abstract

TechViwoEDU Project is addressing the hypothesis that school age children in rural villages, throughout the world, can use advanced low-cost technologies, such as programmable drones, coding robots, and physical & digital manipulative tools for learning and knowledge acquisition. Specifically, they can use these low-cost tools, sometimes partly made from local materials, to acquire developmental skills such as higher-order thinking (problem-solving thinking, critical thinking, rule-based thinking, creative thinking), computational thinking, algorithms, digital and automation technologies, and STEM-based technologies, for future of work and future of jobs. Drones provide a gateway for engaging learners with mathematical topics such as geometric, spatial, topological structures; order structures, temporal structures; and algebraic structures. Drones are also a part of the toolset that enables the students and their coaches to collaborate in active learning involving exploration, discovery, creativity, ingenuity, innovation, competition, and cooperation. A challenge provides effective user interfaces for the comprehension of the huge amount of information that learners must assimilate and accommodate. The project has been launched for the rural community schools of Tsrukpe in the North Dayi District of the Volta Region of Ghana, West Africa, with a student population of about 600, and pilot program for about 60 students.

Keywords: programmable drones, educational drones, educational technology, higher-order thinking, critical thinking, problem-solving, heuristics, computational thinking, STEM, manipulatives, cognitive assistants, cognitive exoskeletons, tangible interfaces, learning virtual dashboard and canvas, educational multi-sided platform, virtual and digital manipulatives, chatbots, LLM, machine learning, AI for children learning, generative AI, digital virtual drones, digital virtual robots, digital twins for education, Ananse stories for education, Ananse stories for STEM learning,

Oware games, mancala games for STEM learning, human intelligence amplification, augmentation, exoskeleton, prostheses

1. Introduction

TechViwoEDU is an ongoing project, targeted at middle school, secondary school and high school age children in rural communities initially in West Africa, but subsequently to be scaled up to the rest of Africa, Asia, and the Americas. The aim of the project is to find and test ways of educating rural children in the context of fast-moving technological trends such as digital technology, automation, digital transformation, robotics, drones, artificial intelligence, cognitive digital assistants and intelligent chatbots.

The guiding vision of the project is that despite the reality of the existence of a digital divide between poor rural communities and urban communities (more realistically considered as digital gap, or even digital chasm), rural children must not be left behind, and do not have to be left behind. Our premise is that there are viable ways to provide for the rural children's active engagement and active learning despite the acknowledged digital gulf between children from resource-limited and those from resource-rich populations and communities.

The chapter is organized as follows: Section 2 discusses the issues and concerns of education in rural village communities; Section 3 focuses on learning by rural village children; Section 4 addresses the educational technology tools and platforms being used in the project; and finally, Section 5 discusses the details of the project under development and implementation.

2. Challenges of STEM education in rural village communities

Rural village communities have the primary characteristic that they are low-resourced, resource-poor communities or very resource challenged. Examples of these scarce resources pertain to modern technologies and include power, energy, Internet, Web, smartphones, and digital platforms. Educational projects can be started in the rural villages, typically using donations and largess of others, although reliance on donations from outside the local community, including members of their own diaspora, will be sporadic and therefore not sustainable.

3. Rural village children

Rural children live in poverty which unfortunately and unavoidably, impacts their preparedness for academic performance. The only minimal assumptions to be made in the development and execution of the project is that (a) they are kids by age; (b) they are growing up in low-resourced or limited-resource environments; (c) their learning deficits and deficiencies can be addressed using digital assistants, digital exoskeletons and learning prostheses, and digital intelligence amplifiers. In a nutshell, these children are deprived of amenities and resources that can help them to be educated about the changing world around them. Our mission is to provide Hope, Future and Promise to the kids through this project, starting with children in three rural schools in Tsrukpe in the North Dayi District of Volta Region in Ghana, West Africa.

4. Educational technology and resources for kids in rural village communities

Educational technology resources currently being incorporated into the TechViwoEDU project include field programmable educational drones, field programmable educational robots, wearables and fitness trackers, physical STEM kits, digital STEM kits, physical manipulatives, digital manipulatives, virtual drones, virtual robots using customization of Chatbot-LLM-Generative AI platforms, and the digital twins of drones and robots (LLM: Large Language Models).

5. The TechViwoEDU project

5.1 Governing principles, hypothesis, assumptions, theoretical foundations

Several theoretical principles underscore the project. The aim is to ensure that identified project goals are achieved. The project should be functionally effective - it is an educational technology in which curriculum and lesson sequences truly support rural village children in achieving their learning objectives and outcomes. A key issue that is being addressed by the project is how to use the educational technology resources effectively to accomplish learning goals and learning objectives. This usage must be Specific, Measurable, Actionable, Realistic, Time-limited (SMART). Finally, the project should be ultra-low cost and sustainable over its operational life. The major underlying principles are briefly described as follows.

1. **Make child learning habit forming:** Build the educational resources and platforms, including curricular development and lesson plan designs to make learning habit forming for children. Here are examples under this principle: Use the MATI-ABI-ORIC framework of the Hook model [1] and the Persuasive technology model [2]. (MATI-ABI-ORIC: Motivation-Ability-Trigger-*-Action-Behavior-Investment-*-Outcome-Result-Reward-Investment-Continuation). For the Motivation (M) and Trigger (T) components, we will *use questions and problems as quests* for knowledge acquisition. The Project expects to derive a lot of the quest problems from the ties of drone applications and robot applications to STEM topics, especially in the subject areas of mathematics. **Figures 1** and **2** indicate concrete areas of applications of flying robots (drones) and robotics, which in general can serve as sources of inspiration for student learner questions, problems, projects and quests for learning STEM.

2. **Technology tools and platforms are safe and secure for children:** Ensure that in the learning environment, the technology tools and platforms are safe and secure for children learners. One of the mottoes in the project is, "Everything that we try to do should always be in the best interest of the children." One way the project has chosen to institutionalize this motto is to specify and build multi-layer encapsulations surrounding the two major avenues by which children learners can access digitally available human knowledge, namely: (a) Internet/Online/Web/Social Media; and (b) Chatbots/LLM/ANN/ML/AI. (ANN: artificial neural-networks; ML: machine learning; AI: artificial intelligence).

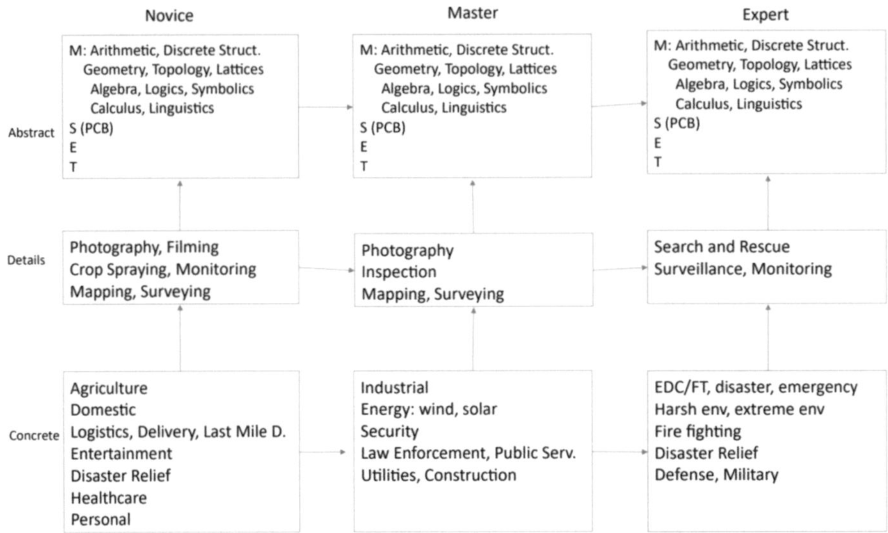

Figure 1.
Application areas of flying robots, drones. [T: Technology → E: Engineering → S (PCB): Science (physics → chemistry, materials → biology) → M: Mathematics].

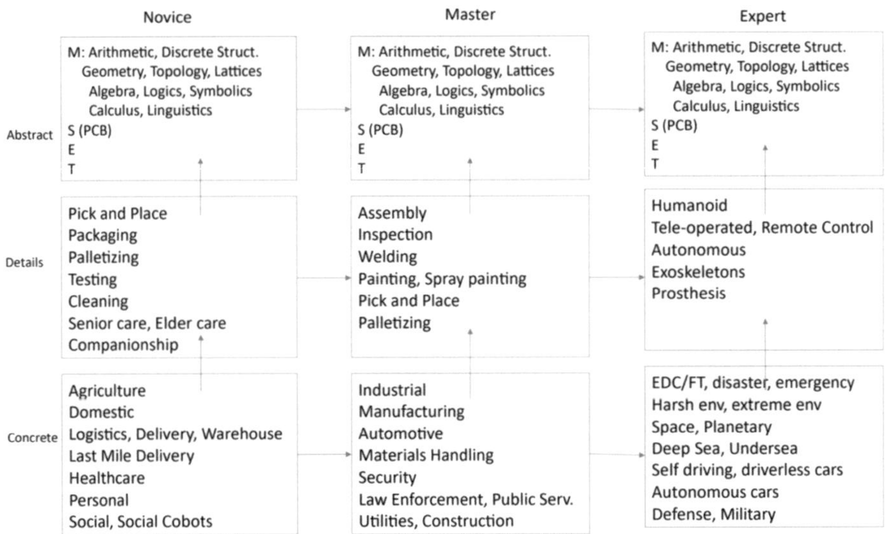

Figure 2.
Application areas of robots, robotics.

The safety and security model is shown in **Figure 3**. The encapsulations are nested gateways, gatekeepers, firewalls, filters and guards. Some of the gateways will serve in roles of prompt engineering and domain specific self-tuning.

It is also intended that the children themselves play active roles in determining and specifying the requirements and characteristics of hygiene, sanitation, and immune system qualities of interfacing the gateway that is used to access and use the knowledge, information, data and intelligence storehouses.

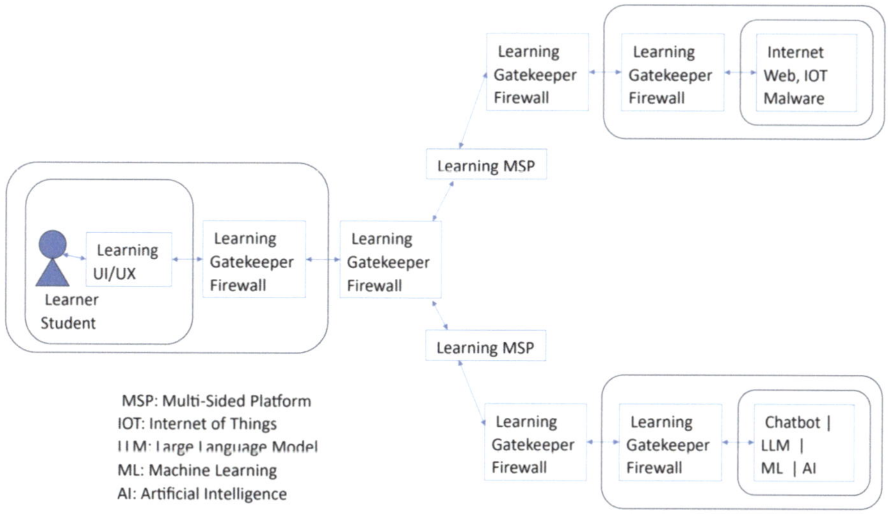

Figure 3.
Multi-layer security and safety filters strategically situated to guard children learners' access of human knowledge storehouses.

3. **Construction-based approaches:** Learning should occur via construction-based approaches including:

    - Creativity for learning;
    - Construction-driven learning, building, making, producing, programming, manipulating things;
    - Exploration, touring, inspection touring, navigation, space and world traversals and navigation;
    - Some learning is executed as epic heroic journeys and adventures, searching through problem spaces, problem-solving spaces (strategies and tactics), and solution spaces.

4. **Augmented, mixed, hybrid reality:** Use augmented, mixed, hybrid reality, superimposing and integrating, blending and mixing physical Edtech resources with digital virtual ones. The project embraces both physical educational drones and robots, as well as AI-based and AI-assisted virtual drones and digital drones, as well as digital twins.

5. **Cognitive, emotional, motivational, and conation aspects:** The Edtech platforms adopted should play the role of being learning exoskeletons, prosthesis, and intelligence amplifying devices for children's learning efforts, including the cognitive, emotional, motivational, and conation and volition aspects.

6. **Active learning:** Ensure that student learners are continuously and persistently immersed and engaged in active learning, instead of just being mere passive consumers of the educational technology. The platform is to be structured and used

so that the student learners are active participants throughout the life cycle of the Edtech platform - from initial functional and non-functional (iLities) specifications and requirements through all phases of evolution and adaptation.

7. **Transition/transformation from novice-master-expert:** Adopt the pedagogical strategy and tactics of regarding students as undergoing a multi-level transition or transformation from novice-master-expert knowledge acquisition, during their learning processes, journeys and engagements. The Games People play model of Child-Parent-Adult framework in [3] supports this strategy, as does the adopting models of learning as evolutionary computation, as well as the concept assimilation and accommodation model in [4] (**Figure 4**).

8. **Rewards:** In the rewards component of the Learning to Habit model, the learning performance assessment is to be based on categories of rewards for self, the hunt and tribe (social group), described in Ref. [1]. This is also related to categorizations of power [5]: physical-coercive-condign-force; financial-monetary-pecuniary, seductive-charismatic-persuasion-influence. Furthermore, some of the project effort is geared towards creating and producing learning materials and processes that can readily generate the DOSE neuro-chemicals in the learners, (DOSE: dopamine + oxytocin + serotonin + endorphins).

9. **Implementation:** Appendix A provides a drill down of more implementation strategies that are being incorporated and integrated into the project.

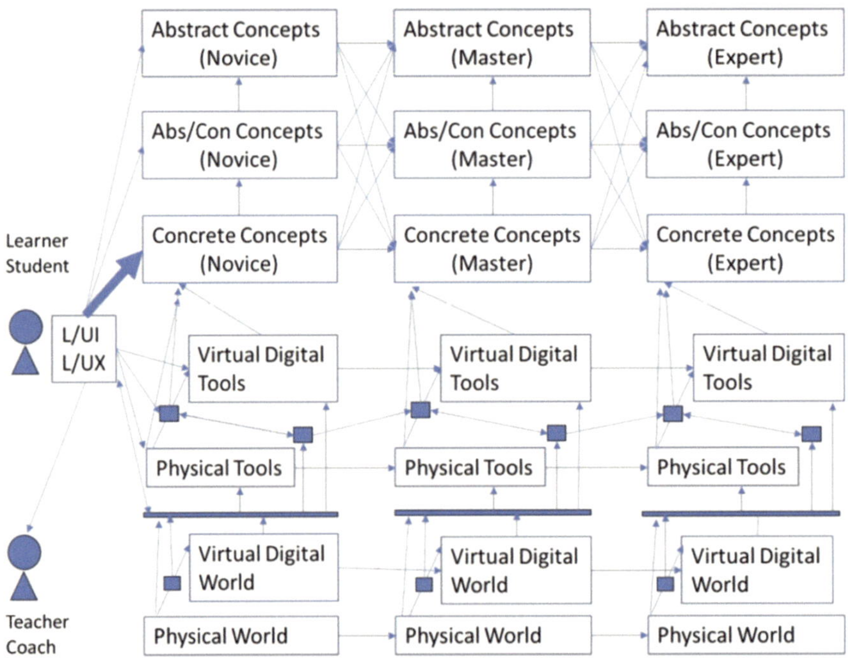

Figure 4.
Learning as novice → master → expert transition life cycle spiral. [L/UI: Learning User Interface; L/UX: Learning User eXperience; Concrete: Focus on sensory-motor + emotion-driven actions, process, manipulations; Abstract: Focus on mental, intellectual, conceptual, theoretical + emotional symbolic processes and manipulations].

5.2 Details of implementation and status

The project is currently in the initial launch-and-takeoff phase. All the relevant forces, resources, capital and stakeholders have been identified. All the requisite groundwork has been made regarding stakeholder awareness, buy-in/ownership, recruitment, persuasion, engagement, involvement, participation, contributions and investments of time, attention and effort.

Our stakeholders include, (a) the children learners or students (the ultimate target customers); (b) the schools, school administration, education staff and faculty; (c) the potential coaching pool, from the local community; (d) the parents, guardians, caretakers and caregivers; (e) rural village community and public including traditional social and modern political leadership, organizations and social groups; (f) government at the national, regional, municipal and local rural village levels; (g) benefactors and friends of the children; and (h) corporate social responsibility actors and directorates.

Identified technology resources are being procured and acquired including: Internet and ICT infrastructure; power and energy from the public utility; educational technology (Edtech) equipment, tools, machines, and devices, including, smartphones, educational drones, educational robots, fitness and health trackers as educational IOT (Internet of Things) devices, STEM kits and manipulatives, origami for education, and access to AI (for education) platforms. Further details are provided in Appendix B.

Students in the initial cohort from three middle schools are learning to use the Scratch programming language [6], to get familiarized with computational thinking, programming, coding, and personal use of digital technology for learning and knowledge acquisition. Scratch is a good choice for a starter computational and programming language, since it is the foundation for many coding and end-user programming languages that are available for physical educational drones and educational robots. The programming platform is also being experimented with to implement some of the computer-based approaches to children's learning that is advocated in [7]. The project sections are also being encouraged to use Origami for Education practices in preparation and anticipation of the drone and robot resources. Extensive work is being done on the development of suitable curricula and activity plans and protocols that can support the learning engagements, using the educational technology tools, vehicles and platforms.

5.3 Discussion

The project implementers are gaining experience, mastery and expertise on how the project can be successfully implemented to meet its vision, mission, and goals. An important consideration which is being addressed from project inception is identifying the strategy that can be used to ensure sustainability of the project. So far, there has been a pleasant surprise, that the project can be potentially fashioned to become a drive or an engine for innovation, in the development and customization of local content for rural villages, that can be commercialized. This will render our project learning centers (PLC) to become sustainable, because they will become profit centers, instead of remaining cost or loss centers.

5.4 Future effort

A planned scaling of the project is to extend the technical, STEM, computational thinking, and higher order thinking project mission to primary school age learners

(K-6/5–9 year olds). The near term future plans for the project also include expansion of coverage to high school (secondary school) age learners, as well as to other regions in Ghana. At the strategic level, the focus will be on continuity, durability, scale, expansion, evolution and continuous improvement. As part of the long term future plans, senior high school (SHS) students in participating schools and clubs will be given the opportunity to become peer mentors to help assemble engineered Edtech resources such as drones and robots, for use by middle school or junior high school (JHS) and primary school (K-3) students. The new generation of Edtech devices can include low-cost local materials such as bamboo, fabrics, and weaving. The high school students will be intimately involved with how the modern AI platforms are customized and specialized in STEM education and learning in rural village communities.

6. Conclusions

The TechViwoEDU initiative is an ongoing project to use educational drones and educational robots, both physical and virtual digital, to act as learning intelligent amplification, augmentation and assistants (IA*), that can support children in rural villages and towns to learn STEM + STEAM + Computational Thinking + Higher Order Thinking (Critical thinking, Problem Solving, Creative Thinking, Rule-guided thinking, + Intuitive thinking + Emotional Intelligence + Social Intelligence).

Acknowledgements

TechViwoEDU Project gratefully acknowledges the acceptance and enthusiastic participation of the headmasters, principals and the ICT specialists at the Tsrukpe secondary and elementary schools: Asiedu Jacob Komla, Aggor Francis, Master Raymond, Bansa Coura, and Bediaku Wisdom. The Project is also grateful for the faculty and staff of the schools, who are willing to get trained to become educational coaches for the Project. Furthermore, the Project is making rapid progress because of contributions of ideas, guidance and discussions of several people: Aseye Gadagoe, Gabriel Dedu, Bernice Heloo, Christine Gbeckor-Kove, Yao Ababio, and Michael Afenya. To all of them, Ayeekoo! (Thank You!)

Conflict of interest

The authors declare no conflict of interest.

A. Appendix

Some other learning approaches that the Project is incorporating into the curriculum design and activity planning include the following ideas, especially for learning action and behavior investment (ABI) components.

a. Use the W*H* (Who, Whom, What, Which, Why, When, Where, How, How soon, How much, How often, etc.), as well as the 5 Whys model (ask why five times in sequence about a topic), and the Thematic-semantic case model + Separation of Concerns, Challenges, Aspects and Roles (SOCAR).

b. Use game-like quest, models focusing on problems and questions for exploration, touring, inspection, problem solving, creativity (Create-Read-Update-Delete (CRUD), social cooperation and collaboration, competition, contests, epic heroic journeys in game-like models in [8, 9], as well as elaborated by [10].

c. Also, use the Bloom educational tasks taxonomy model [11].

d. Further, use the PDCA cycle and (CPI, Agile, Kaizen, Toyota Way) approach to learning [12].

e. Use Entity-Relationship-Attribute-Value-Domain (ER model, ERAVD model) [13], which provides and foundational compositional framework for the construction-based learning approaches of [4, 14–17], as well as for concept map, schema, frame based and case based reasoning models for learning and knowledge acquisition. It also provides a suitable medium for using the Algebra approach to learning: Algebra = Collections, (data) Structures+ Manipulations, Operations+Rules-Laws-Axioms-Identities. Examples of such algebras are abstract data types (ADTs) in CS.

f. Use Heuristic Problem-solving model in [18], the Decode-Encode-Solve-Check (DESC) model, and the continuation models described in [19–21].

g. Incorporate modern heuristic techniques inspired by AI and machine learning. These are techniques currently grouped under natural computing and soft computing, including, evolutionary computing (Generation-Of-Diversity (G.O.D)-Evaluation-Selection), Darwinism and also as in [3]; search, tabu search, rule-based (logical-crisp and fuzzy) techniques, probabilistic statistical techniques.

B. Appendix

Initial resource items include, (new, used and refurbished versions of):
Educational drones
Educational robots
Coding drones, Coding robots
Coding toys
Fitness trackers & Health trackers
Batteries (AA and AAA)
Charging adaptors
Power strips
Smartphones
Tablets
MiFi cellular signal booster devices
Internet access equipment
Internet access subscriptions
Access to Chatbots/LLM, including Claude/Anthropic, ChatGPT/OpenAI, Bard/Google

Access to Scratch App for Programming and Coding

Initial sequence of engagement, by educ resource types:

Stage-1: Origami in the Classroom, Origami clubs, Origami camps.

Stage-2: Educational blocks, educational kits, manipulatives, construction kits, Montessori kits, play kits, User eXperience (UX) card decks for storytelling, narrative and story creation. Problem-solving via student creation of Ananse stories and storytelling whose themes are problem-solving quests, and concept learning using Oware and Mancala game boards.

Stage-3: Coding toys, STEM kits, STEAM kits.

Stage-4: Coding, programmable educational drones and robots.

Author details

John-Thones Amenyo[1*] and Wolanyo Kpo[2]

1 City University of New York, United States

2 Chicago State University, United States

*Address all correspondence to: jtamenyo@york.cuny.edu

References

[1] Nir E. Hooked: How to Build Habit-Forming Products. New York: Portfolio Penguin; 2014

[2] Fogg BJ. Persuasive Technology. Boston, MG: Morgan Kaufman Publishers; 2003

[3] Berne E. Games People Play: The Psychology of Human Relationships. New York: Grove Press; 1964

[4] Wadsworth BJ. Piaget's Theory of Cognitive and Affective Development: Foundations of Constructivism. White Plains, NY: Longman Publishers; 1996

[5] Galbraith KJ. The Anatomy of Power. Boston, MA: Houghton Mifflin; 1983

[6] Resnick M, Maloney J, et al. Scratch: Programming for all. Communications of the ACM. 2009;**52**(11):60-67

[7] Schank RC, Cleary C. Engines for Education. Hillsdale, NJ: Lawrence Erlbaum; 1995

[8] Kim AJ. Game Thinking. Burlingame, CA: gamethinking.io; 2018

[9] Lazzaro N. Why We Play Games: Four Keys to More Emotion without Story. Oakland, CA: XEODesign, Inc.; 2004

[10] Campbell J. The Hero's Journey. Novato, CA: New World Library; 2014

[11] Anderson LW, Krathwohl DR, Bloom BS. A Taxonomy for Learning, Teaching, and Assessing: A Revision of Bloom's Taxonomy of Educational Objectives. Complete ed. London, UK: Pearson Longman; 2001

[12] Deming EW. Out of the Crisis. Boston, MA: MIT Press; 2000

[13] Chen PPS. The entity-relationship model-toward a unified view of data. ACM Transactions on Database Systems. 1976;**1**(1):9-36

[14] Papert S. Mindstorms—Children, Computers and Powerful Ideas. New York: Basic Books; 1980

[15] Papert S. The children's Machine: Rethinking School in the Age of the Computer. New York: Basic Books; 1993

[16] Gutek GL. The Montessori Method: The Origins of an Educational Innovation: Including an Abridged and Annotated Edition of Maria Montessori's The Montessori Method. Lanham, MD: Rowman & Littlefield; 2004

[17] Froebel F. The Student's Froebel Adapted from Die Erziehung Der Menschheit von F. Creative Media Partners: Fröbel; 2016

[18] Polya G. How to Solve It. Princeton, NJ: Princeton University Press; 2014

[19] Lakatos I. Proofs and Refutations, The Logic of Mathematical Discovery. Cambridge, UK: Cambridge University Press; 1976

[20] Popper KR. Conjectures and Refutations: The Growth of Scientific Knowledge. New York, NY: Basic Books; 1962

[21] Poincaré H. Science and Method. Reprint. North Chelmsford, MA: Courier Corporation; 2013

Chapter 9

Aerial Drones for Fire Disaster Response

Ramasenderan Narendran, Thiruchelvam Vinesh,
Soon Hou Cheong and Han Xiang Yee

Abstract

The significance of fire in human society encompasses essential functions like illumination, warmth, and cooking but also poses immense risk when uncontrolled, leading to catastrophic damage and loss of life. Traditional firefighting responses are often hindered by geographical and logistical challenges, resulting in delays that exacerbate the severity of fires. This research introduces an innovative solution through the use of an autonomous firefighting drone, designed for round-the-clock surveillance and rapid response to fire scenes. Utilizing image processing and neural networks, the drone can efficiently detect fire and smoke, serving as the first responder, and is equipped with fire extinguishing balls to initiate suppression. The work extends to explore the application of AI edge aerial drones in disaster response, not only to fires but also floods and landslides, particularly in Malaysia and Southeast Asia. By focusing on various urban, peri-urban, and rural contexts, the research delineates potential implementation strategies aimed at enhancing situational awareness for first responders and reducing response time to reach victims, thereby facilitating more effective disaster response operations. The study's findings point to a considerable advancement in firefighting technology that could lead to decreased fire damage and saved lives, filling a critical gap in the disaster response playbook. This advancement in firefighting technology enhances response times, decreases fire damage, and ultimately, saves lives.

Keywords: fire, fire extinguishing ball, fire fighting, artificial intelligence, unmanned aerial vehicle

1. Introduction

The ongoing evolution of cities into the scientific, economic, administrative, and cultural epicenters of their respective nations has been a remarkable phenomenon of the twentieth century. This evolution, despite delivering substantial improvements in living conditions, also precipitates a host of challenges. These include increased traffic congestion, which stresses infrastructure while amplifying air and noise pollution, a housing crunch prompting cities to expand both horizontally and vertically, and issues such as water scarcity and waste management.

Increasingly complex urban infrastructure has amplified concerns regarding safety and security. In response, communities have implemented various technical approaches to create a multidisciplinary emergency rescue team which comprises of first responders, paramedics, firefighting personnel who are ready to be called in upon request, for energy and communication systems.

The past century has witnessed the evolution of new building materials, construction styles, and methods of resource utilization in urban settings. However, despite significant progress in fire prevention, fire incidents in urban areas continue to grow more frequent, complex, and hazardous, posing major challenges to first responders.

The Department of Statistics Malaysia (DOSM) reported [1] 50,720 fire incidents in 2019, a 24.1% increase from 2015. Amidst the restrictions during Covid-19, the reporting of 15,393 fire incidents from March to August, saw nearly half of which were due to preventable open burning. The smoke from these fires not only harms the environment but also adversely impacts human respiratory systems. Structural fires, mainly ignited by electrical faults (49.1% of cases during the MCO), gas leaks (17%), sparks (14%), and flammable items such as lighters, candles, and matches (7–2%) contribute significantly to fire incidents [2].

While geography shields Malaysia from natural disasters like earthquakes and tsunamis, the most common threats stem from man-made incidents, including traffic accidents, fires, and floods. Fire, one of the essential elements from ancient times, represents power and intensity. However, when uncontrolled, fire can be deadly. The damage caused by fires in various settings, including residential buildings, vehicles, and even airplane engines, inspires fear and devastation from afar. Yet, the reality experienced by those at the centre of such incidents is vastly more distressing and often marked by terror, suffering, and tragically, fatal outcomes.

Fire spreads quickly, causing extensive damage and leaving lasting physical and psychological scars on its victims. Firefighting involves combating this relentless element with the forceful application of water—a task carried out by brave fire brigades. However, their courage is accompanied by substantial risks. Firefighters often face unknown variables, such as the overall structural stability, fire origin, potential for collapse, building temperature, smoke density, and more. They plunge into hazardous environments, sometimes blind or with minimal briefing, where unpredictable circumstances can lead to fatalities. The world [3] sees more than 50 firefighter deaths annually, excluding the tragic loss of 340 firefighters in the World Trade Centre disaster.

Efforts to reduce the risks faced by firefighters have been made worldwide, with technology taking the lead. Unmanned firefighting machines equipped with capabilities for monitoring, inspecting, and extinguishing fires have been developed as a safer alternative. However, a vital yet often overlooked factor is the pre-evaluation of the fire environment. Information about fire intensity, smoke concentration, the number of trapped individuals, and the presence of explosive materials remains unknown until firefighters arrive at the scene. Hence, thorough pre-evaluation can enhance safety and efficiency in firefighting operations.

Unmanned Aerial Vehicles (UAVs) are now employed for surveillance in various sectors, including wildlife management, fire behavior monitoring, and package delivery, in many [4] countries. However, in Malaysia, the adoption of UAVs in firefighting is minimal, primarily due to local humidity and high temperatures. Currently, drones require human operators for guidance and face constraints such as distance limitation and battery life. They also need precise navigation and rapid response to avoid obstacles. 3D area mapping could be a solution, allowing drones to follow a predetermined path and provide valuable information about the fire's external environment. However, high-resolution 3D imagery necessitates sophisticated and expensive vision cameras, which can inflate costs when implemented in a drone swarm. Furthermore, the risk of damage to drones remains high due to various factors that can cause failure.

Despite advancements in technology, fire incidents and the associated fatalities continue to rise. The National Fire Protection Agency (NFPA) suggests an ideal response time of 9 minutes and 20 seconds, of which includes 14.3% of turnout time and 85.7% of travel time [5]. Kamarulzam Malik Abdullah, the director of the Sabah Fire and Rescue Department, cites factors such as a shortage of fire stations and geographical constraints for this delay.

The deployment of autonomous firefighting drones and fire extinguisher balls can reduce response time and help firefighters gain easier access to the incident scene.

The potential use case for UAVs remains high as the enable both accessibility and data collection for a multitude of analytics [6] which can be harnessed for timely and critical information which equips fire fighters with situational awareness to combat the disaster successfully with minimized human casualties and property damage.

2. Fire detection

An innovative approach [7] to fire detection systems through the use of triboelectric nanogenerators (TENGs). These devices, proposed initially by Wang's group in 2012, have the ability to harness energy from the wind, thus offering a potential solution for fire detection in power failure scenarios common in fire accidents.

TENGs offer certain distinct advantages over conventional methods. For one, they have a high sensitivity to the frequency of mechanical vibrations. They are also smaller in size compared to traditional turbines, allowing for greater space utilization.

In the event of a fire, power sources are often cut off to prevent further complications such as short circuits or power leakage. However, this also means that remote rescue electronic devices cease functioning. This is where TENGs come into play. They can serve as an intelligent fire alarm system that continues to operate during a fire accident.

TENGs can also harness wind energy even in low-speed regions and convert it to useful power. They offer a novel power source alternative to conventional ones like power plants and batteries, also reducing the length of wiring needed from these traditional sources. Additionally, the use of TENGs optimizes space utilization.

However, the previous iterations of TENGs relied on elastomers and wind, operating through fluid-induced vibration (FIV). This operation subjected the elastomer to complicated stress and continual fatigue fractures, thereby affecting the device's efficiency. Since wind usually flows at slow speeds and in casual directions around the Earth's surface, the device required low base wind velocity and multidimensional wind energy for efficient data collection in fire detection.

To overcome these issues, Zhang et al. proposed a TENG based on fluid-induced vibration, referred to as F-TENG. The F-TENG consists of six identical TENG units, spinning switches, and a low-cost lever mechanism. This design enables the device to persistently collect and analyze wind energy from any given direction. The incorporation of six identical TENG units and spinning switches facilitates gathering wind energy from multiple angles, lowers the starting speed, and reduces the risk of deformation to the TENG. This approach offers a more efficient solution for fire detection compared to the conventional F-TENG that uses an elastomer.

Furthermore, the researchers have developed a methodology to detect fires based on the concentration of smoke. This involves using a capacitor and a rectifier bridge driven by the F-TENG. This not only facilitates the detection of fire but also allows for the calculation of the direction and speed of fire spread.

More impressively, the system can anticipate potential fire hazards by utilizing the data gathered by the F-TENG, such as humidity, temperature, wind speed, and wind direction. This predictive feature enhances the system's potential as a robust fire prevention and safety mechanism.

In recent research, a blockchain technology is applied to develop [8] a novel fire detection system that incorporates image processing. The innovative approach involves the use of a "blockchain golden fox," which consists of various interconnected units. The fundamental unit within this system, known as the meta-network, is constructed from a blend of activation functions and connection methods.

In this unique setup, neurons are connected with protrusions, employing both linear and non-linear mapping as activation functions. These functions subsequently confine the neuron's output to a specific range.

To activate the alarm system, a weight value derived from the blockchain is utilized. This is achieved using the term frequency-inverse document frequency (TF-IDF) algorithm, as the weight value resides between 0 and 1.

During the process of image processing, the image is first converted into grayscale. This is due to the fact that red color, which is associated with fire, has a higher wavelength compared to green and blue color. By making the image grayscale, the red color or light is made more visible, thereby enabling faster and easier fire detection.

Following the grayscale conversion, the image then undergoes binarization to enhance the contrast of the video or image being inputted into the system. Turning the colored or grayscale image into binary is also beneficial in reducing pixel interference.

The image processing concludes with the morphological process, which serves to eliminate or reduce noise in the image. This includes the erosion and dilation process, further enhancing the efficiency and accuracy of the fire detection system.

In the research [8] conducted, a comprehensive image processing workflow, was used to detect fire based on blockchain technology. The process commences with image pre-processing, which reduces the quantity of image data and facilitates valuable data discovery. Subsequently, image segmentation takes place, partitioning the image into distinct segments, each with their unique properties, thereby highlighting areas of interest. The creation of a feature map, which assigns values and extracts vital data from the image, ensues. Image matching, the final step, entails comparing these extracted features with a mapping table for object or fire recognition.

Throughout the experimentation process, it was noted that high temperatures triggered a fire alarm response within 13.3 seconds, while high smoke density levels elicited a response in 18.1 seconds.

In a separate study [9] image processing involving color filtering and histograms was used for fire detection. An examination of RGB (red, green, and blue) and HSV (hue, saturation, and value) color spaces showed that filtering in RGB color spaces was more suitable for detecting fire in static light conditions, such as interiors, compared to open areas. However, for image processing, the threshold technique in HSV color spaces was more advantageous, as demonstrated by their previous work [10]. This study also involved color filtering, beginning with an image conversion from

RGB to HSV. The HSV image was subsequently transformed into grayscale and binary formats for better visibility. Their experimental setup involved placing four different colors (yellow, blue, green, and red) next to the fire, which were not detected as the HSV value was set to detect fire color. The optimum parameters for detecting fire were established as hue ranging from 0–16 to 0–20, saturation from 74–164 to 74–168, and value from 200–228 to 200–232.

In the application of drones for fire detection, it was discovered that the onboard camera could only detect fire if the drone was flying at an altitude not exceeding 8 meters above the fire.

A technique [7] used flame emission spectroscopy in conjunction with image processing for fire detection, noting the technique's superior response time compared to gas sensors and thermal sensors. This method also allowed for a more nuanced understanding of the combustion process by tracking various oxidants. In terms of image processing, the YCbCr color space was used, with a temporal smoothing algorithm improving image quality and enhancing fire detection sensitivity.

Furthermore, it is recommended [11] the integration of the Internet of Things (IoT) with image processing for fire detection. While they acknowledged the widespread use of the RGB color model due to its simplicity and applicability in various applications, they noted its limitations in color recognition and object description. Comparatively, the HSI color space was easier for humans to understand and better at non-uniform illumination. However, this color space was unstable due to its angular nature. The YCbCr color space excelled in reducing image size and improving image quality but was dependent on the original RGB signal. A satellite network [11] and unmanned aerial vehicles (UAVs) is utilized to collect images for fire detection, supplemented by sensor nodes placed in smart cities to gather environmental data such as humidity, temperature, light intensity, and smoke.

2.1 Fire size detection

In their fire sizing study, a novel approach towards active fire detection (AFD) based on the interpretation of Landsat-8 Imagery via Deep Multiple Kernel Learning was explored. The research utilized a recent dataset developed by De Almeida et al., containing image patches of 256 × 256 pixels that illustrate wildfires occurring in various locations across all continents. The dataset thereby provides an opportunity to adapt the MultiScale-Net to a diverse set of geographical, climatic, atmospheric, and illumination conditions [12].

The method of Deep Multiple Kernel Learning is highlighted as being particularly effective for spectral-spatial feature extraction from remote sensing images. To address the limited availability of training samples, the researchers employed a straightforward data augmentation technique, which led to the generation of 27 different configurations in this study [12].

The research group also introduced a novel indicator, the Active Fire Index (AFI), devised to enhance the accuracy of thermal and spectral band analysis. AFI draws its efficacy from the high sensitivity of the Landsat-8 sensor to fire radiation, and it is derived from the SWIR2 and Blue bands [12].

Active Fire Index (AFI):

$$AFI = \frac{\rho_7}{\rho_2} \tag{1}$$

where ρ_7 and ρ_2 represent the SWIR2 and Blue values in Landsat-8 images, respectively.

Eq. (1) can be used to calculate AFI, where ρ_7 and ρ_2 represent the SWIR2 and Blue values in Landsat-8 images, respectively. AFI's utility lies in its ability to distinguish fires from their background owing to the high reflectance in the SWIR2 spectrum and the relatively low reflectance in the Blue spectrum [12]. It further proves effective in eliminating smoke and possible clouds present in the image scenes, which are frequent hurdles in active fire detection.

However, the AFI method does not come without its limitations, for instance, it can underperform when non-fire objects that are bright have a high reflectance in the SWIR2 and a low reflectance in the Blue spectra. Furthermore, the accuracy of fire detection can be compromised due to certain inherent characteristics of fire such as the limited spatial resolution of satellite images and the significant influence of the Earth's dense atmosphere [12].

The independence of the proposed method from the thermal bands paves the way for future studies to explore the potential of Sentinel-2 data for AFD, which could offer higher spatial and temporal resolution. Moreover, the research team suggests that the processing time of the proposed method could be evaluated using cloud platforms like Google Earth Engine (GEE) [12].

Another novel approach to understand and model wildfires was using a method to isolate individual fires from the commonly used moderate resolution burned region data [13].

In their study, they acknowledge the necessity to separate individual fires from large clusters of burned area as these fires are produced by extensive burn patches. This comes as a challenge in current global fire systems, as the available satellite data products only identify active fire pixels. Hence, they developed the Global Fire Atlas, which outlines individual fires based on a fresh methodology that identifies the direction of fire spread, the position and timing of individual fire regions and calculates the size of the fire, its period, daily expansion, fire line length and speed [14].

This Global Fire Atlas algorithm was applied to the MCD64A1 Col.6 burned-area dataset [14], and it was found that the minimum detected fire size is one MODIS pixel, equivalent to approximately 250 m × 250 m. This result provides a very precise and promising avenue for future research.

In another research endeavor, a comprehensive study evaluating the influence of the loss function, architecture, and image type for Deep Learning-Based Wildfire Segmentation [5]. They processed two image types, namely the visible and the Fire-Gan fused, and evaluated 36 resultant combinations of the architectures and loss functions with reference to three different metrics—MCC, F1 Score, and HAF [15].

Their analysis demonstrated that the Akhloufi + Dice + Visible combination yields the best results for all metrics. Moreover, the Akhloufi architecture and the Focal Tversky loss function were found to be prevalent in the top five for all metrics. Thus, considering the performance evaluation and correlation analysis, the combination of Akhloufi + Focal Tversky + visible was recognized as the most robust performer due to its consistent results with minimal variance [15].

3. Fire fighting strategies

The firefighting drone functions as a first responder at the fire scene, capable of warning people in the area and notifying the fire department. Moreover, it comes with a feature that enables it to extinguish parts of the fire, creating pathways for people to escape and providing clear paths for firefighters to enter [16].

A system [16] that can deploy a fire extinguishing ball to assist in wildfire fighting. The mechanism for payload deployment was constructed with components such as a motor, power supply, and receiver, allowing the drone to receive signals from the user to open the valve through the motor and release the fire extinguishing ball.

The researchers found that the 0.5 kg weight of the fire extinguishing ball was insufficient in extinguishing fires effectively. The choice of the 0.5 kg weight during the experiment was due to budget constraints, and it took approximately 3–4 seconds to activate the fire extinguishing ball, covering an area of about a meter in diameter. However, the researchers believe that a larger fire extinguishing ball could extinguish fires over a larger area. According to the official webpage, the founder mentioned that the 1.3 kg weight of the fire extinguishing ball can cover an area up to 3 cubic meters [16].

The YOLOv4 algorithms [17] were improved by reducing the neural network's weight using the Kernelized Correlation Filter (KCF) algorithm for tracking and adaptive learning rate template update when tracking fails. The results were obtained from a multi-rotor copter by testing the adaptive tracking strategy. KCF used a non-linear kernel function to handle non-linearity and non-stationarity and adapted the threshold value for the Average Percentage of Correctly Estimated (APCE) metric to provide a more accurate evaluation of the tracking algorithm's performance [17].

The use of YOLOv5 for object detection and combined it with the DeepSORT algorithm for tracking and positioning objects using UAVs. DeepSORT is a real-time object tracking algorithm that uses a deep neural network-based object detector and a simple online and real-time tracking algorithm, SORT (Simple Online and Realtime Tracking), to effectively track objects even in challenging situations [18].

The nearest neighbor (NN) method [19] to measure and track selected objects. They employed the nearly constant velocity (NCV) motion model for discrete-time kinematic modeling in object tracking. The NCV model assumed that the target's velocity was constant over a short period, and the Kalman filter was used to predict and estimate the target's position and velocity at each time step [19]. The nearly constant velocity (NCV) motion model is a common choice for the discrete-time kinematic model for a target in object tracking. This model assumes that the target's velocity is constant over a short period of time, and that the position of the target at each time step is determined by its position and velocity at the previous time step. In the NCV model, the target's position and velocity at time step k are represented by the state vector x_k = [x_k, y_k, vx_k, vy_k]^T, where x_k and y_k are the target's position coordinates, and vx_k and vy_k are its velocity components. The state transition function for the NCV model is given by:

Nearly constant velocity (NCV):

$$x\_k = Fx\_k - 1 + w\_k - 1 \tag{2}$$

where F is the state transition matrix, and w_k−1 is the process noise. The state transition matrix F is typically set to the identity matrix with the velocity components set to the time step between frames. The Kalman filter algorithm is then used to estimate the target's position and velocity at each time step by using the state transition function and the measurement function. The measurement function is used to relate the target's state vector to its measured position in the current frame. The filter uses the measurement and prediction to correct the estimate of the state vector. The NCV motion model is simple to implement and computationally efficient, making it well-suited for real-time tracking applications. However, it does not consider the acceleration of the target, which can lead to errors in the estimated position and velocity when the target's motion is not truly constant.

A method for recognizing the self-location of a drone flying in an indoor environment using ultra-wide band communication module DWM1000 was proposed [20]. The self-localization algorithm uses trilateration and the gradient descent method to determine the drone's position with an error of within 10–20 cm. Real-time 3D position information of the drone can be obtained and used for autonomous flight control through a deep learning-based control scheme. This scheme improves upon a conventional CNN algorithm by incorporating deeper layers and appropriate dropouts in the CNN structure, which uses input data from a single camera. When a drone experiences drift, it may not move in a straight line but instead veer slightly to one side. In this situation, the real-time first-person view (FPV) of the drone is used as input data in the proposed CNN structure to predict the pulse-width modulation (PWM) value for the roll direction, which helps to correct the drift and keep the drone moving in the desired direction. The goal of this work was to create a drone that could safely navigate in environments where GPS signals are not available, such as tunnels and underground parking garages. To achieve this, a control board based on a digital signal processor and ultra-wideband modules were developed, along with control algorithms for stable flight. A 3D position estimation algorithm using a proposed 2D projection method was implemented to determine the drone's position, which was then used for hovering motion control and way-point navigation.

4. 3D area mapping

There will be ways for performing 3D mapping using a 2D LiDAR instead of 3D LiDAR sensor. Using LiDAR Lite v2 sensor can be detecting objects up to 40 m. It was low cost and high performances. Due to its lightweight characteristic, it was suitable for UAV. Using the servo motor as the actuator module for turning the LiDAR sensor. The Arduino UNO board was used as the microcontroller that moves the servo motors. Upon detection, a point cloud graph will be plotted since the ultrasonic sensor was a distance-based sensor. For turning it into a 3D scan, different step will be used. Different steps of the servo motors will provide different resolution of images. A chair was used in the research method. It concludes that, the smaller the steps needed, the longer the processed time needed which have more scanned points to be plotted.

There were many applications can be done using 3D mapping. In this case, it was used for urban analysis for both spatial distribution of population and residential buildings. These data called census data can be collected by urban area units of a dimension. Unfortunately, the geometry of these census tracts was found changing between two censuses, either on spatial aggregation or by spatial disaggregation. This happened due to new structures being built or when urban densification occurred. Using dasymetric mapping with different geometrical shapes and different types of information such as lost spaces, public spaces, proximity, and urban density become every useful. Dasymetric mapping was a spatial interpolation technique that allows the re-allocation of area data from source to targets geometries. In order to operates this condition for a lower cost, using a UAV will be a low-cost -friendly technology.

UAV offers great potential in operating airborne sensor system and computer vision for monitoring 3D buildings and spatial planning. Swinglet CAM was a fixed-system system that used for acquiring 3D point cloud over the region. With the help of few sensors such as RGB sensors for georeferenced images pair, capturing real time points using GNSS systems and internal measurement unit (IMU). Some of the parameters can be obtained like position of laser sensors based on one or more GNSS base station. IMU can provides parameters such as sensor's altitude and heading angles (roll, pitch, or yaw) Moreover, LiDAR or satellite imagery can be the tools for extracting 3D points clouds. Information of the mapped building such as height, area, volumetric information can be extracted using 3DEBP (3D extraction building parameters).

UAV point cloud can be obtained by multi-stereo image matching processing of eight ground control points allowing a set of 3D points being estimated through those stereo pair pixels. There will be millions of points being captured so these points have been filtered through clustering large applications (CLARA) algorithm. Large dataset such as point cloud will be suitable to use CLARA because due to its partitioning ability to divide these data into few sub-groups.

Even though there were mean error for the estimated building block volume when referencing from original dataset. In the early days, the monitoring method for fires event were using satellites system as it was high efficiency, accurate and automatic operations [21]. However, this method were not reliable and poor signals from the satellite if was in remote areas. There were many alternatives in terms of fire monitoring which can be used which was the remote sensing which had advantage in large area monitoring and high spatial resolution images were efficient. Comparing to InSAR techniques that provides area surveying and mapping had an advantage since it can monitor day/night with different weather conditions but on the bad side, it was limited to its corresponding time as fire monitoring required continuous observations. With satellite imaging, the probability of false alarms was high as well as dense fog, cloudy and rainy images may affect its monitoring.

One of the methods was using UAVs as the replacement of satellite images since UAVs can provided real time monitoring at low altitude and fast acquisition of data. Its reliable and simple to operate. Attaching a digital camera will be ideal as it can captured basic surface model and regular images for analysis. Other options will be mounting a multiple lens camera for proper texture information of the mapped area. This can show the strength of UAVs in comparing against the traditional method of satellite imaging. By implementing image processing algorithm into the UAVs, drone mapping can be done with the help of data analytics called ThingSpeak cloud platform. It utilized the platform and provided capabilities of accessing data both online and offline. Using Drone deploy software, it can be able to map drones which offers web and app-based platform. It can provide high resolution images and videos to target area.

With the help of 3D mapping with SLAM, even rover at mars and moon can estimate its current positions and construct a map for surrounding. Some of the sample sensors that can used were RGB-D and LiDAR to provide high dense 3D point cloud but limited communication links, data storage, power supply could be a problem. Alternatives, a stereo camera, or monocular vision could work due to lightweight and lower power consumptions. The implementation of Visual SLAM mapping can be performing using a pair of stereo images used to train the deep learning model, the disparity map was estimated. This map was initially used to create 3D point cloud.

A good implementation of SLAM technique for terrain perceptions and mapping results were determines by the choice of sensors and sensor fusion. According to the research paper [22], different methods were used. First, it uses monocular SLAM framework with extended Kalman filter (EKF) that able to track unconstrained motion of a rover but for this to work, those distinct feature must be well distributed. The biggest problem of this method will be scale ambiguity and measurement drift since there were no inertial and range sensor when using single camera. Alternatives, data from RGB-D SLAM will acquire both depth per-pixel and visual texture information. However, lighting circumstances can have interferences causing noisy and homogenous point-clouds. Some of the researcher proposed LiDAR SLAM as it can solved the robustness problems as the research paper had mentioned the global terrain map was made with sparse-method and batch of alignment algorithm [22] LiDAR camera fusion with SLAM had great advantages since it can perfectly make use of these both sensors individually. Other than that, used of stereo SLAM was also one of the proposed methods because it can create a terrain map through sensor fusion and using the Gaussian-based Odometer error model, it can improve the accuracy by predicting the non-systematic error from wheel interaction.

To prove the concept, training dataset on earth was not practical since in this paper, the focus was on mars and moon. A self-supervised deep learning model was used and 2D and 3D convolution layer based on geometric-based CNN were trained using the images in the subset collected by the pairs of stereo cameras. The 2D convolution layer will extract images features from each image and constructed a cost volume. Then, the 3D convolution layer will be aggregating to infer the disparity value using probability distribution for each pixel. Lastly, the disparity map was constructed as a regressed from the probability distribution.

According to the research done, SLAM mapping using combination of single-scan Terrestrial Laser Scanning (TLS) point cloud and Mobile Laser Scanning (MLS) point cloud can addressed the consistency problem and maintained accuracy mapping without using the GNSS-IMU system. LiDAR Odometry and global optimization will be the main key to solve the problem. The LiDAR odometry will estimate the motion of each detected frame of MLS point cloud relative to single-scan TLS data. Real and virtual features will be extracted as a part of tree stem mapping using NDT algorithm. Virtual features were sample that reconstructed the tree stem centrelines while real features were evenly sample the point cloud. Layer clustering method will be used to extracted virtual features where else Differences of Gaussian (DOG) method will be used to extract real features. In terms of global optimization, it mitigates the accumulative error and transforms all points cloud into common coordinate system. By adopting a method based on global map that does not consider loop closure and the adjustment of all data to address the global optimization of SLAM.

This research was done to present the optimization of SLAM using Gmapping algorithm Different SLAM algorithm had been used and invented such as Gmapping, Hector SLAM and Karto SLAM. Standards dataset and ground truth map the parameters from source code, C++ program were used to optimize the parameters in a different way. 2 ways were used which were optimizing parameters separately and see its results or the second optimization where parameters collectively and see its results [23].

First method will be optimizing the parameters separately by their functions. The parameters chosen will be editable which means user can provide new values and run the programmed again to conduct evaluation compared to dataset. The main parameters were changed uniformly in equity manner which was uniform optimization and second was references where time between successive recalculations of the map which was the map update interval. Gmapping algorithm were used in this method. After optimizing, map results and necessary significant results were varying in each different parameters testing.

Optimizing parameters separately can led to a little bit of changes on the map relatively to the dataset. Although the results look similar to each other even though we optimize the parameters separately, but it does not mean that the result will be an uniform development as all it were changing were one single parameters [23].

In the journal article title "Mapping and 3D modeling using quadrotor drone and GIS software" currently the main obstacle in 3D mapping would be the cost of data acquisition for high resolution satellite imagery especially for mapping in weekly or daily basis. Researcher proposed an alternative will be using UAV which was low cost, had a high-resolution image and can acquire any time with limited restrictions that needed to comply [24]. In this journal article paper, it approached the problem with developing 3D models by photogrammetric data taken using drone quadcopters. In order to acquire high resolution images, it uses industrial drone, the DJI Inspire 2 drone supported with high quality camera with dreadlocks so that images taken were stable and suitable for mapping activities. The industrial camera was professional as it was the first aerial camera that capable of recording lossless 4 K video in RAW with framerates up to 30 fps and average bitrate of 1.7 Gbps. With the use of powerful Micro Four Thirds Sensors (MFT), camera that have can achieved high image quality, light and compactness make it great for recording anywhere anytime. It also captures stunning 20MP images with details with stabilization dreadlocks integrated with 3-axis guarding level. OF course, with such high specs' camera recording, the specification of PC must be top notch as well. Software used was the AgiSoft Metashape, a software that performed photogrammetric processing of digital images and generates 3D spatial data to be used in GIS applications.

SLAM can be used to create a map for an unknown terrain by estimating the positions of the obstacle from the created map. The Visual Simultaneous Localization and Mapping (VSLAM) method can be achieved by using a monocular camera. To achieve a comparison and had a better result, the two main method of VSLAM can be used which was the direct-based method and features-based method. These two methods can be done by extracting information from the images taken by the camera. From the journal, Large Scale Direct (LSD-SLAM) was preferred for the direct based method as when implementing the LSD-SLAM, it used the whole images as input and point cloud were built up based on the images pixel while for feature-based method, the preferred choice was the ORB-SLAM where data was extracted from a orb and the representation of pixels will be recorded [25]. In terms of accuracy, the direct-based method will be much more accurate as more points will be captured and recorded compared the feature-based method.

As mentioned in this section, 3D area mapping with SLAM would a very good way as it reconstructs the terrain and mapped unknown terrain. There were many different types of SLAM but VSLAM was one of the famous types as it can directly acquire points from the images taken by monocular camera. Direct method and indirect method can be used to differentiate these two approaches to implement this SLAM. Direct method using the LSDO-SLAM can acquired better results as it uses the whole images as input and directly acquired different density pixel for map plotting. Direct method often be the better choice when the external surrounding had poor lighting and poor texture surfaced. Comparing to direct method, the indirect method, the ORB-SLAM were used and by extracting some of the features for the observations of the 3D map.

5. Conclusion

The exploration of drone systems as detailed in this chapter presents a multifaceted analysis of contemporary technology with broad applications. While the capabilities of these drone systems have been found to deliver numerous benefits in terms of safety and effectiveness, it has become evident that there are intrinsic limitations that present challenges to their optimal use.

1. Battery power: the first significant limitation is a shortage of battery power, resulting in limited flight time. This not only restricts the drone's operational capabilities but also affects the overall efficiency. The constant need for battery replacement or recharging poses logistical challenges.

2. Payload capacity and motor power: linked to the issue of battery power is the drone's limited payload capacity and insufficient motor power. These constraints have been found to restrict the ability to lift heavier objects or carry additional equipment, such as larger batteries. The consequent low lift capacity hampers the drone's maneuverability, especially under adverse weather conditions or heavy payload.

3. System high temperature: the incorporation of 4G (LTE) cellular connectivity via Raspberry Pi, while enhancing communication, has resulted in high system temperatures due to the absence of a cooling mechanism. This heat issue threatens the system's stability and longevity, risking failures and reduced lifespan.

4. Camera performance: the last major limitation noted is the suboptimal performance of the drone's camera, delivering lower-resolution photos and video streams. This affects the drone's efficiency in tasks requiring high-definition imagery, such as aerial surveillance or inspection.

However, these limitations are not insurmountable, and the chapter has also provided a pathway towards mitigating these challenges:

1. Improving battery configuration: investigating alternative power sources, including high-capacity batteries or solar charging, could substantially extend flight duration.

2. Enhancing motor power: a comprehensive analysis to identify suitable motor settings could significantly improve lift capacity and maneuverability.

3. Addressing high-temperature issues: including efficient cooling mechanisms, such as heat sinks or cooling fans, would ensure system stability and longevity.

4. Upgrading camera module: investing in higher-quality cameras or gimbal-stabilized camera systems could lead to better-quality imagery.

The integration of these recommendations requires a systematic approach, including in-depth research, extensive testing, and due consideration of compatibility and integration challenges. By addressing these limitations, the drone system could realize its full potential, becoming more versatile and efficient in various applications, such as surveillance or inspection. This chapter's insights and recommendations contribute to the broader discourse on drone technology, setting a constructive path for future innovations and implementations. It lays the groundwork for researchers, practitioners, and policymakers to leverage drone technology, harnessing its strengths while thoughtfully addressing its weaknesses.

Author details

Ramasenderan Narendran*, Thiruchelvam Vinesh, Soon Hou Cheong and Han Xiang Yee
Asia Pacific University, Kuala Lumpur, Malaysia

*Address all correspondence to: narendran@apu.edu.my

References

[1] Nasir NA, Rijal NS. Fire safety management systems at commercial building: Pasir Puteh supermarket. e-Proceeding. 50 p

[2] Timbuong J. Firefighters dealt with more fires in first MCO. The Star Newspaper. 2020

[3] Fahy RF, LeBlanc PR, Mollis JL. Firefighter Fatalities in the United States-2010. Quincy, MA: National Fire Protection Association; 2011

[4] Tan CF, Dhar MS. Fire Fighting Mobile Robot: State of the Art and Recent Development. Australian Journal of Basic and Applied Sciences. 2013;7:220-230

[5] Averill J, Moore-Merrell L, Ranellone R Jr, Weinschenk C, Taylor N, Goldstein R, et al. Report on High-Rise Fireground Field Experiments. Technical Note (NIST TN). Gaithersburg, MD: National Institute of Standards and Technology; 2013. DOI: 10.6028/NIST.TN.1797

[6] Ramasesenderan N, Rajasekaran T, Sivanesan S. Analysis and optimisation of lubrication viscosity on the bearing operations. In: Journal of Engineering Science and Technology, Special Issue on SIET2022, May, 51-57. Kuala Lumpur, Malaysia: Asia Pacific University; 2022

[7] Qiu X, Xi T, Sun D, Zhang E, Li C, Peng Y, et al. Fire detection algorithm combined with image processing and flame emission spectroscopy. Fire Technology. 2018;**54**(5):1249-1263. DOI: 10.1007/s10694-018-0727-x

[8] Zhao F. Application research of image processing technology for fire detection and fire alarm based on blockchain. Mobile Information Systems. 2022;**2022**:1-11. DOI: 10.1155/2022/9304991

[9] Ahmad A, Prakash O, Khare K, Bhambhu L. Detection of flames in video by color matching and energy separation. Journal of Real-Time Image Processing. 2019;**16**(5):1625-1642

[10] Ahmad A, Khare K, Bhambhu L. Detection of flames in video by color matching. IEEE Sensors Journal. 2017;**17**(6):1795-1802

[11] Amit K, Rajat C, Tripti S, Bhaskar T. A review of fire detection systems with new results on image processing for early stage forest fire detection. Wireless Networks. 2020;**26**:4577-4595. DOI: 10.1007/s11276-019-02125-7

[12] Rostami A, Shah-Hosseini R, Asgari S, Zarei A, Aghdami-Nia M, Homayouni S. Active fire detection from landsat-8 imagery using deep multiple kernel learning. Remote Sensing [Internet] 2022;**14**(4):992. DOI: 10.3390/rs14040992

[13] Andela N, Morton DC, Giglio L, Paugam R, Chen Y, Hantson S, et al. The Global Fire Atlas of individual fire size, duration, speed and direction. Earth System Science Data. 2019;**11**:529-552. DOI: 10.5194/essd-11-529-2019

[14] Louis G, Luigi B, David PR, Michael LH, et al. The collection 6 MODIS burned area mapping algorithm and product. Remote Sensing of Environment. 2018;**217**:72-85. DOI: 10.1016/j.rse.2018.08.005. Available from: https://www.sciencedirect.com/science/article/pii/S0034425718303705. ISSN 0034-4257

[15] Ciprián-Sánchez JF, Ochoa-Ruiz G, Rossi L, Morandini F. Assessing the

impact of the loss function, architecture and image type for deep learning-based wildfire segmentation. Applied Sciences. [Internet]. 30 Jul 2021;**11**(15):7046. DOI: 10.3390/app11157046

[16] Aydin B, Selvi E, Tao J, Starek MJ. Use of fire-extinguishing balls for a conceptual system of drone-assisted wildfire fighting. Drones. 12 Feb 2019;**3**(1):17

[17] Tian J, Shen L, Zheng Y. Genetic algorithm based approach for multi-UAV cooperative reconnaissance mission planning problem. In: Proceedings of the International Symposium on Methodologies for Intelligent Systems; 2006 Sep 27-29; Bari, Italy. pp. 101-110

[18] Wu W, Liu H, Li L, Long Y, Wang X, Wang Z, et al. Application of local fully Convolutional Neural Network combined with YOLO v5 algorithm in small target detection of remote sensing image. PLoS ONE. 2021;**16**(10):e0259283. DOI: 10.1371/journal.pone.0259283

[19] Yeom SK, Seegerer P, Lapuschkin S, Binder A, Wiedemann S, Müller KR, et al. Pruning by explaining: A novel criterion for deep neural network pruning. Pattern Recognition. 2021;**115**:107899. DOI: 10.1016/j.patcog.2021.107899

[20] Son SJ, Kim HS, Kim DH. Waypoint and autonomous flying control of an indoor drone for GPS-denied environments. IAES International Journal of Robotics and Automation (IJRA). 2022;**11**(3):233-249. DOI: 10.11591/ijra.v11i3.pp233-249

[21] Amit S, Pradeep KS, Yugal K. An integrated fire detection system using IoT and image processing technique for smart cities. Sustainable Cities and Society. 2020;**61**. DOI: 10.1016/j.scs.2020.102332

[22] Sungchul H, Antyanta B, Jae-Min P, Minseong C, Hyu-Soung S. Visual SLAM-based robotic mapping method for planetary construction. Sensors (Basel). 2021;**21**(22):7715. DOI: 10.3390/s21227715

[23] Gunaza Teame W, Zhongmin Prof W, Yu Dr Y. Optimization of SLAM Gmapping based on Simulation. International Journal of Engineering Research & Technology (IJERT). 2020;**9**(04). DOI: 10.17577/IJERTV9IS040107

[24] Budiharto W, Edy I, Jarot SS, Andry C, Heri N, Alexandar Agung Santoso G. Mapping and 3D modelling using quadrotor drone and GIS software. Journal of Big Data. 2021;**8**(48). DOI: 10.1186/s40537-021-00436-8

[25] Krul S, Pantos C, Frangulea M, Valente J. Visual SLAM for indoor livestock and farming using a small drone with a monocular camera: A feasibility study. Drones. 2021;**5**:41. DOI: 10.3390/drones5020041

Chapter 10

Depth Learning Methods for Bridges Inspection Using UAV

Hicham Sekkati and Jean-Francois Lapointe

Abstract

This paper is investigating learning methods using depth as a cue measurement that can be used for bridge inspection. We investigate learning methods based on mono, stereo, and multiview image input and discuss the constraints that allow some methods to perform better than others in various scenarios. We go over the state-of-the-art deep learning methods, including supervised and unsupervised methods. These methods will be compared and evaluated, based on constraints, performance, and accuracy, and how top methods should be selected for each scenario. The same database should be used for fair comparison between all methods ensuring that evaluations are unbiased, replicable, and meaningful.

Keywords: depth, 3D reconstruction, deep learning, bridge inspection, UAV

1. Introduction

Tragedies such as the recent collapse of the Morandi bridge in Italy [1] remind us of the importance of good and regular bridge inspections. Such inspections are conducted mainly manually but the advent of new technologies such as drones and depth estimation using deep learning paradigms has the potential to automate part of the task. Depth map estimation for bridge inspection can provide valuable information about the three-dimensional structure of the bridge. It allows for the identification of surface irregularities, cracks, deformations, and overall conditions. This process provides valuable information for analyzing the structural integrity and identifying potential issues. Several techniques can be used for depth map estimation in bridge inspection, including Time-of-Flight (ToF) cameras, Structured Light scanning (SL), Laser scanning or Light Detection and Ranging (LiDAR) systems, as well as photogrammetry-based depth estimation. ToF cameras emit infrared light and measure the time it takes for the light to bounce back from the bridge surface. This information is used to estimate the distance to each point on the surface, creating a depth map. ToF cameras can provide real-time depth information, making them suitable for dynamic inspections. SL scanning involves projecting a pattern of light onto the bridge surface and capturing the deformed pattern using a camera. By analyzing the distortions in the pattern, depth information can be calculated. This technique is effective for capturing detailed depth maps of bridge surfaces and can be performed using handheld devices or mounted on vehicles or drones. LiDAR technology can also be utilized for depth map estimation, by emitting laser pulses and measuring their time of flight, LiDAR scanners can generate accurate depth information of the bridge's surface. High-resolution LiDAR scanners can capture detailed depth maps, facilitating precise analysis of the bridge structure. In photogrammetry-based techniques, and by employing computer vision algorithms, depth maps can be estimated from regular images captured by Unmanned Aerial Vehicles (UAVs) or drones during bridge inspections. Various techniques, such as depth from single image, structure from motion (SfM), or multi-view stereo, can be employed to extract depth information from the image data.

Each of the above techniques has its advantages and limitations, and the choice depends on factors such as the desired level of accuracy, resolution, portability, and budget. It is advisable to consult with experts in the field of bridge inspection or 3D imaging to determine the most suitable depth estimation method for a specific bridge inspection project. However, this paper focuses on photogrammetry-based methods and specifically the last advances using deep learning techniques to generate depth maps from images.

This paper first talks about depth perception and then discuss various ways of obtaining depth, be it by using pictorial cues, from monocular video, or from stereo and multi-view.

2. Depth perception

Perception refers to the ability to interpret and organize stimuli from the surrounding environment, enabling effective understanding and behavior. The visual system plays a crucial role as one of the primary sources of stimuli for human beings. It comprises more than one million axons from each eye, responsible for capturing light reflected by objects. Research on human perception suggests that the visual system utilizes multiple sources of information to comprehend and infer the depth structure of scenes. The human visual system relies on various monocular or binocular cues present in two-dimensional retinal images to gather information that helps in perceiving the depth of the scene. Monocular cues can be divided into two categories: pictorial cues and motion-based cues.

Pictorial cues, or image cues, are derived from visual features observed in a static view of a scene. The most common pictorial cues used in computer vision methods for depth estimation from a single image are texture variations [2], shading [3], and defocus [4]. Texture variations are translated such as objects that are closer to the viewer tend to exhibit more detailed and distinct textures, while objects that are farther away appear to have less detailed or blurred textures. This texture gradient helps us infer depth. The distribution of the direction of edges or lines in a scene changes as objects recede into the distance. The spacing between these lines becomes smaller as objects get farther away, giving us a sense of depth. Depth from shading is a technique used to estimate the depth or 3D structure of a scene based on the shading or variations in brightness and contrast within an image. It relies on the principle that the distribution of light and shadows on objects can provide valuable information about their shape and depth. Depth from defocus is a depth estimation technique that utilizes the blur or defocus information in an image to infer the distance of objects. These algorithms take advantage of the fact that objects in the focus plane of a camera appear sharper, while objects that are out of focus exhibit varying degrees of blur. By analyzing the amount of blur in an image, these algorithms can infer the relative depth of different objects in the scene. A good survey on methods using depth from defocus can be found in ref. [5]. Let us remind that depth from defocus technique is fundamentally different from depth from focus in the sense that the later uses a stack of images to model the blur in image while the former technique uses a single image. The stack of images can be obtained by varying the camera aperture like in ref. [6] or the focal length like in refs. [7–11]. In the next section, we only review learning depth from a single image.

On the other hand, motion-based cues make use of observer motion and leverage motion parallax, that is, nearby objects appear to move faster in the retinal image compared to distant objects. In contrast, binocular cues rely on the perception of depth through disparities between two different viewpoints of the same scene. By comparing the differences in the views from each eye, the brain can accurately triangulate the distance to an object. Binocular cues offer a high level of precision in estimating distances. The aim of the following sections is to go over the last advanced research on deep learning techniques for each category to estimate depth.

3. Depth from pictorial cues

Deep learning methods have been successfully applied to estimate depth from a single image, leveraging the power of neural networks to learn complex mappings between image features and depth information. The first deep-learning method to estimate depth from a single image was proposed in ref. [12]. Image cues are learned as multi-scale features. The method uses two-step process involving two deep neural networks to predict depth information for a given scene. The first step is performed by a coarse-scale network. This network takes an input image as its input and predicts the depth of the scene at a global level. The second step involves a fine-scale network. This network takes the coarse depth map (output of the coarse-scale network) and refines it within local regions. The method achieves state-of-the-art results on both NYU Depth [13] and KITTI [14] datasets. The authors in ref. [15] proposed a framework to model the conditional probability on depth with conditional random field (CRF) and learn the probability distributions using deep convolutional neural network (CNN). The method has outperformed the classical methods on both indoor and outdoor scenes using both the public datasets NYU depth and the Make3D range image [16]. In ref. [17], the method also uses two CNN to capture both global and local scales while jointly estimating depth and semantic segmentation from a single image. The method in ref. [18] has trained a CNN to learn the relative depth ordering between pairs of points in the image. The same network was trained to learn independently the reflectance and shading in the image, however, no interaction between these metrics was taken into account. A better structural relationship between points in the image was learned by a CNN in ref. [19]. This method involves training a neural network to characterize the local geometry of a scene by predicting depth derivatives of various orders, orientations, and scales at every image location. In ref. [20], a method that combines a CNN and regression forest was presented to regress depth in the continuous domain. In ref. [21], the authors proposed a fully convolutional architecture (ResNet) for depth prediction enabling the generation of dense output maps with higher resolution, while significantly reducing the number of parameters required. Furthermore, the model can be trained using one-tenth of the data compared to the previous state-of-the-art approaches. An improvement of the previous method's accuracy was presented in ref. [22] by applying a post-processing *via* fully-connected conditional random fields (CRF). More improvements using CRFs in cascades was presented in ref. [23]. In ref. [24], two cascade-deep fully connected CNNs were proposed to learn both global and local feature maps that are propagated to estimate depth. Most methods learn depth as a regression model and code implicit structure of the scene with CNNs features, but in ref. [25], a method was presented that explicitly modeled the defocus blur in an image and link it to image depth. In ref. [26], a method was presented that learns depth from defocus, unfortunately only qualitative results on NYU depth dataset were shown. Quantitative comparisons with state-of-arts learning methods on this dataset were not reported. **Tables 1** and **2** summarize the evaluation of depth estimation from state-of-the-art pictorial-based methods using both Make3D [16] and NYU Depth [13] datasets, respectively.

Method		**Error (C_1)**			**Error (C_2)**	
	AbsRel	**log10**	**RMS**	**AbsRel**	**log10**	**RMS**
Saxena et al. [2]	—	—	—	0.370	—	—
Roy et al. [20]	—	—	—	0.260	**0.119**	**12.400**
Liu et al. [15]	0.314	0.119	8.600	0.307	0.125	12.890
Anwar et al. [24]	0.213	0.075	**2.560**	**0.202**	**0.312**	**0.079**
Laina et al. [21]	0.176	0.072	4.460	—	—	—
Xu et al. [23]	**0.184**	**0.065**	**4.380**	**0.198**	**4.530**	**8.560**

Table 1.
Result comparisons of depth evaluation from pictorial-based methods on the Make3D dataset. Best performance is marked with bold fonts.

Method		**Error**			**Accuracy**	
	AbsRel	**log10**	**RMS**	$\delta < 1.25$	$\delta < 1.25^2$	$\delta < 1.25^3$
Zoran et al. [18]	0.400	0.420	1.200	—	—	—
Liu et al. [15]	0.230	0.095	0.824	0.614	0.883	0.971
Wang et al. [17]	0.220	0.094	0.745	0.605	0.890	0.970
Eigen et al. [12]	0.215	0.285	0.907	0.611	0.887	0.971
Roy et al. [20]	0.187	0.078	0.744	—	—	—
Chakrabarti et al. [19]	0.149	0.205	0.620	0.806	0.958	0.987
Cao et al. [22]	0.141	0.060	0.540	0.819	0.965	0.992
Laina et al. [21]	0.127	0.055	0.573	0.811	0.953	0.988
Xu et al. [23]	0.121	0.052	0.586	0.811	0.954	0.987
Anwar et al. [24]	0.094	0.039	0.347	—	—	—
Carvalho et al. [25]	**0.036**	**0.016**	**0.144**	**0.993**	**1.000**	**1.000**

Table 2.
Result comparisons of depth evaluation from pictorial-based methods on the NYU depth dataset. Best performance is marked with bold fonts.

3.1 Loss functions

Learning depth from a single image is a challenging task due to the inherent ambiguity. However, there are techniques that leverage deep learning models to estimate depth from a single image. When training such models, a common choice for the loss function is the depth regression loss, which measures the difference between the predicted depth map and the ground truth depth map. One popular loss function for depth regression is the mean squared error (MSE) loss, given by:

$$L_1 = \sum_{x,y} (Z(x,y) - Z^\star(x,y))^2 \tag{1}$$

where $Z^\star(x,y)$ is the depth map predicted by the model and $Z(x,y)$ is the ground truth depth map. The MSE loss penalizes large errors between the predicted and ground truth depth values. Minimizing this loss helps the model learn to estimate depth accurately. Alternatively, you can also use other variations of the loss function such as the Huber loss or the smooth L1 loss, which provide a balance between the absolute and square losses and can be less sensitive to outliers. These loss functions can be advantageous when dealing with noisy or sparse depth measurements. When training models to learn depth from a single image, additional constraints or regularization terms might be necessary to improve the quality of the estimated depth. Some common techniques include incorporating geometric or semantic information, enforcing local smoothness, or using multi-scale depth supervision. Overall, the choice of the loss function depends on the specific requirements and characteristics of depth estimation task. Experimentation with different loss functions and regularization techniques can help find the most suitable approach for different applications.

4. Depth from monocular video

Estimating 3D interpretation from a monocular video is a fundamental and challenging topic in visual perception. Two common techniques used for this purpose are structure from motion (SfM) and simultaneous localization and mapping (SLAM). In the context of monocular video, SfM involves jointly estimating the camera motion and depth map of the scene, while SLAM involves jointly estimating the camera trajectory and the 3D structure of the scene. Monocular depth prediction using pairs of frames or more can be particularly challenging. This is because it requires reasoning about the relative camera pose, as well as estimating the disparity or optical flow between the frames. Furthermore, there is an inherent ambiguity in scale when using only monocular input, unless additional information or a consistent SLAM reconstruction pipeline is employed. The relative camera pose estimation is crucial for understanding the spatial relationship between frames and is necessary to compute accurate depth maps. Determining the camera motion accurately becomes more difficult when dealing with larger displacements, occlusions, or scene dynamics. Incorrectly estimated camera poses can lead to inaccurate depth predictions. Additionally, estimating disparity or optical flow between frames is challenging due to factors such as textureless regions, occlusions, and large displacements. These factors can introduce errors and ambiguities in the depth estimation process. Moreover, when using only monocular input, there is an inherent scale ambiguity. That is, without additional information, it is challenging to determine the absolute scale of the scene, leading to depth maps that are only accurate up to an unknown scale factor. Despite the difficulties, researchers continue to develop methods that leverage monocular depth prediction using pairs of frames. These methods often combine deep learning techniques with geometric constraints and SLAM-like approaches to improve the accuracy of depth estimation and mitigate the inherent challenges. Ongoing research in this area aims to push the boundaries of monocular depth prediction and address the inherent limitations of single-camera input.

4.1 Supervised deep learning methods

Supervised deep learning methods have made significant progress in addressing the problem of determining 3D interpretation from monocular video. By training neural networks on large-scale annotated datasets, these methods can learn to estimate depth, motion, and other geometric properties from single-camera input. One popular approach to estimate depth is to use convolutional neural networks (CNNs) applied independently at each frame of the video. These networks take an image as input, as seen in the previous section, and output a depth map that represents the scene's 3D structure. By leveraging the large amounts of labeled data, CNNs can learn to infer depth cues such as perspective, texture gradients, and occlusion patterns. A common dataset used to compare methods in this category is KITTI [14]. In ref. [27], given a pair of frames and camera intrinsics, a deep architecture, computes depth, 3D camera motion, a set of 3D rotations and translations for the dynamic objects in the scene, and corresponding pixel assignment masks. However, the method uses a single image deep architecture for depth estimation.

Additionally, recurrent neural networks (RNNs) and particularly the convolutional LSTM networks have been employed to capture temporal dependencies and motion information in video sequences. By incorporating temporal context into the learning process, these models can estimate not only depth but also camera motion, object motion, and scene dynamics. In ref. [28], the proposed ConvLSTM network learns depth maps from a set of N consecutive video frames in a depth-supervised setting, allowing the ConvLSTM network to perform spatiotemporal reasoning about the image-depth map relationship.

To further enhance performance, supervised methods often make use of additional cues, such as optical flow or semantic segmentation. Optical flow provides dense pixel-level motion information [29], which can aid in-depth estimation and object tracking. Semantic segmentation helps in understanding the scene's layout and can guide the depth estimation process by leveraging object boundaries and semantic context.

However, it is important to note that despite the advancements, challenges remain in accurately determining 3D interpretation from monocular video. Factors such as occlusions, lighting variations, and scene complexity can still pose difficulties for supervised methods. Nonetheless, ongoing research and the continuous development of more sophisticated deep learning architectures hold promise for further improvements in tackling this problem.

4.1.1 Supervised loss function

When learning supervised depth from motion, a common approach is to use a loss function that compares the predicted depth map with the ground truth depth map. One such loss function is the photometric loss, which measures the difference between the rendered image using the predicted depth map and the actual input image. The photometric loss can be defined as follows:

$$L_1 = \sum_{x,y} (I_1(x,y) - I_2(x,y))^2 \tag{2}$$

where I_1 is the original input image and I_2 is the image rendered using the predicted depth map and camera parameters, which can be done using techniques such as differentiable warping or inverse depth warping. In addition to the photometric loss, you can also incorporate smoothness regularization to encourage smooth depth predictions. The smoothness loss penalizes large depth gradients and helps produce more visually coherent depth maps. One common smoothness regularization term is the total variation loss, which can be defined as follows:

$$L_2 = \sum_{x,y} \|\nabla Z\| \tag{3}$$

where $|\nabla Z\|$ is the gradient of the predicted depth map in the x and y directions. The total loss for learning depth from motion can be a combination of the photometric loss and the smoothness regularization term:

$$L = L_1 + \alpha L_2 \tag{4}$$

where α is a weighting factor that controls the relative importance of the photometric loss and the smoothness regularization term. By minimizing this total loss using techniques like gradient descent, you can train a model to learn depth from motion. Keep in mind that this is just one possible approach, and depending on your specific requirements and constraints, you may need to modify or customize the loss function accordingly.

4.2 Unsupervised deep learning methods

While supervised deep learning methods have achieved notable progress in determining 3D interpretation from monocular video, unsupervised deep learning methods have also shown promise in tackling this problem. Unsupervised approaches aim to learn 3D representations from unlabeled or self-supervised data, eliminating the need for costly manual annotations. One popular technique in unsupervised learning is based on the concept of "self-supervision." By leveraging the temporal coherence of consecutive video frames, unsupervised methods can learn representations that capture the underlying 3D structure of the scene. These methods often utilize techniques such as photometric consistency, geometric consistency, or depth and ego-motion prediction.

In *photometric consistency-based* methods, the network learns to generate a synthesized view of the input frame from a different viewpoint using estimated depth or motion. The consistency between the synthesized view and the actual input frame is maximized during training, encouraging the network to learn meaningful depth representations. In ref. [30], the photometric consistency was achieved by training the network in a manner analogous to an autoencoder.

Geometric consistency-based methods exploit the geometric relationship between frames. They aim to minimize the disparity or reprojection error between multiple views of the same scene. By leveraging geometric constraints, the network can learn to estimate depth and camera motion. The method in ref. [31] takes into account this geometry by learning camera pose between each two frames.

Depth and ego-motion prediction methods train the network to directly predict depth maps or motion vectors from single images or consecutive frames. These predictions are compared to ground truth or photometrically warped frames, respectively, to supervise the learning process. In ref. [32], photometric consistency was taken into account in training step while 3D geometry consistency was achieved by reconstructing 3D points cloud from depth and directly comparing the points cloud in a common reference frame. Nonrigid motion of dynamic objects in the scene was taken into account in ref. [33] by adding ResFlowNet architecture [34]. Photometric and geometric consistencies were combined in ref. [35] in a way to minimize the discrepancy between the reconstructed optical flow obtained from depth and egomotion, and the optical flow generated using FlowNet [36]. Dynamic scenes were handled in ref. [37] by learning objects motion independently from the egomotion without an explicit motion segmentation. Likewise, the motion model of moving objects in the work [38] is tackled by optical flow estimation using view synthesis objective as supervision, again with the assumption of photometric consistency. In ref. [39], a method was presented by adding another term that explicitly segments the scene into competing background and foreground masks. In most unsupervised methods, including mono or stereo-SfM (Structure from Motion) approaches, photometric consistency is a crucial principle used to guide the learning process. Photometric consistency is based on the assumption that the appearance of the same point in different views should remain consistent under different camera poses. In the context of monocular or stereo video, this consistency is expressed using a warping function and is often referred to as the view-synthesis loss [31]. In ref. [40], a generalization of the photometric loss was used by coupling the spatiotemporal variations in image sequence to the scene geometry with the goal to supervise both camera motion and depth in a new learning framework. **Table 3** summarizes the evaluation of depth estimation from state-of-the-art monocular motion-based methods using KITTI [14] dataset.

Method		**Error**				**Accuracy δ**		
		AbsRel	**SqRel**	**RMS**	**RMSlog**	**<1.25**	**$<1.25^2$**	**$<1.25^3$**
Supervised	Vijayanarasimhan et al. [27]	—	**0.770**	—	—	—	—	—
	Kumar et al. [28]	**0.137**	**1.019**	**5.187**	**0.218**	**0.809**	**0.928**	**0.971**
Unsupervised	Zhou et al. [31]	0.208	1.768	6.856	0.283	0.678	0.885	0.957
	Garg et al. [30]	0.169	1.080	5.104	0.273	0.740	0.904	0.962
	Mahjourian et al. [32]	0.163	1.240	6.220	0.250	0.762	0.916	0.968
	Yin et al. [33]	0.155	1.296	5.857	0.233	0.793	0.931	0.973
	Zou et al. [35]	0.150	1.124	5.507	0.223	0.806	0.933	0.973
	Casser et al.(M) [37]	0.141	1.026	5.291	0.215	0.816	0.9452	0.979
	Rajan et al. [39]	0.140	1.070	5.326	0.217	0.826	0.941	0.975
	Sekkati et al. [40]	0.137	**0.947**	**5.019**	0.216	0.838	0.933	0.970
	Chen et al. [38]	**0.135**	1.070	5.230	**0.210**	**0.841**	**0.948**	**0.980**

Table 3.
Result comparisons of depth evaluation from monocular motion-based methods on the KITTI dataset. Best performance is marked with bold fonts.

By training on vast amounts of unlabeled video data, unsupervised methods can capture 3D cues and learn to interpret depth, motion, and other scene properties. They can also handle challenging scenarios, such as dynamic scenes, textureless regions, and varying lighting conditions. While unsupervised methods have shown promising results, there are still challenges to overcome. The quality and accuracy of the learned representations heavily depend on the design of the self-supervision tasks and the data distribution. Additionally, the unsupervised learning process can be computationally expensive and may require substantial computational resources. Nevertheless, the development of unsupervised deep learning methods for 3D interpretation from monocular video holds great potential for advancing our understanding of the 3D world and enabling applications in robotics, augmented reality, and autonomous systems.

4.2.1 Unsupervised loss functions

Let us consider two nonconsecutive frames from the image sequence $I(x, y, t)$ at time t_1 and t_2, denoted by $I_1 = I(x, y, t_1)$ and $I_2 = I(x, y, t_2)$, respectively. The general idea followed by a previous work [41], and subsequently by others, is to minimize the photometric loss generated by the image difference:

$$L_1 = \sum_{x,y} |I_1(x, y) - I_2(\tau(x, y))| \tag{5}$$

where τ is the warping function that maps pixel from I_2 to I_1. Using image warping and image similarity metrics to supervise learning frameworks has certain limitations, particularly when dealing with large baseline views, occlusions, and image gradients. While these approaches can be effective in many cases, they may not fully capture the complexities of the scene geometry and structural edges, leading to some shortcomings. Several methods have been proposed to address these limitations, but they may not always be explicitly related to scene geometry. For example, image similarity metrics can help guide the learning process, but they might not directly capture the underlying scene geometry. Estimating depth and understanding the 3D structure of the scene is inherently related to scene geometry, which involves estimating accurate depths and surface orientations. Simple image similarity metrics may not fully encapsulate these geometric properties.

Now let us consider two consecutive frames of the image sequence $I(x, y, t)$ at times t and $t + 1$. We denote the spatiotemporal derivatives of the image sequence by (I_x, I_y, I_t). Then, the 3D brightness constraint for rigid objects' motion can be expressed by ref. [42]

$$\Gamma(\mathbf{T}, \boldsymbol{\omega}, Z) = I_t + \mathbf{s} \cdot \frac{\mathbf{T}}{Z} + \mathbf{q} \cdot \boldsymbol{\omega} = 0 \tag{6}$$

where $\mathbf{s}$ and $\mathbf{q}$ are two quantities expressed in terms of image gradients and camera intrinsic parameters. Then, the problem of learning jointly the depth Z and egomotion, parameterized by translational and rotational motions $(\mathbf{T}, \boldsymbol{\omega})$, can be stated as the following loss minimization

$$L_2 = \sum_{x,y} |\Gamma(\mathbf{T}, \boldsymbol{\omega}, Z)| + \mu |\nabla Z| \tag{7}$$

where the first term reduces the loss when the prediction deviates from the 3D brightness constraint, and the second term stands for smoothing depth to avoid both overfitting and the trivial null solution. Minimization of the loss L_2 will overcome all the above shortcomings related to minimizing L_1 and by adding other constraints instead.

5. Depth from stereo and multi-view

Deep learning methods have been widely employed for depth estimation from stereo or multiview images. These methods leverage convolutional neural networks (CNNs) to learn the mapping between input image pairs or sets and their corresponding depth maps. DispNet is a popular deep-learning architecture specifically designed for stereo depth estimation. It consists of a CNN-based encoder-decoder network that takes a stereo pair of images as input and predicts a dense disparity map, which can be converted to depth. The network is trained using a supervised learning framework with ground truth depth maps. Pyramid Stereo Matching Network (PSMNet) is another deep learning architecture for stereo depth estimation. It introduces a spatial pyramid pooling module to capture multi-scale information and a stacked hourglass network structure to refine the disparity estimation. PSMNet has demonstrated excellent performance in stereo depth estimation tasks. MC-CNN is a deep learning method that takes advantage of multiple views of a scene to estimate depth. It takes a set of calibrated images as input and processes them through a shared CNN architecture to predict the depth map. MC-CNN exploits the inter-view geometric relationships to improve depth estimation accuracy. Generative Adversarial Networks (GANs) have also been utilized for depth estimation from stereo or multiview images. GAN-based methods often involve training a generator network to generate depth maps from input images and a discriminator network to distinguish between real and synthesized depth maps. This adversarial training helps improve the quality and realism of the predicted depth maps. The disparity estimation method in [41] uses a CNN network for computing matching distances between image patches followed by a cross-based aggregation to compute the disparity map. In ref. [43], a CNN was trained in a supervised way to estimate disparity between stereo images from stereo video datasets. An implementation in GPU was presented in ref. [44] to learn feature correspondences faster. In ref. [45], stereo matching is enhanced using conditional random fields (CRF) to improve the accuracy and coherence of the depth estimates. CRF Is a pro”abil'stic graphical model that models the dependencies between variables in a structured manner. In the context of Semi-Global Matching (SGM) [46], the spatial-variant penalty parameters were learned by regularization terms applied to the disparity map to enforce smoothness and coherence. SGM employs a penalization approach where the disparity differences between neighboring pixels are penalized and controlled by the penalty parameters. In ref. [47], a CNN method with differentiable layers was presented that learns an end-to-end mapping from an image pair to disparity map. A refinement by adding another CNN stage was presented in ref. [48].

In ref. [49], a method was presented to train a CNN network that performs end-to-end unsupervised depth estimation with a training loss that enforces left–right depth consistency inside the network. Similarly, the method in ref. [50] learns self-supervised stereo matching as finding the disparity map that best warps between the stereo image pair. In ref. [51], CNN architecture was proposed to jointly unsupervise

Method		Error					Accuracy δ		
		AbsRel	**SqRel**	**RMS**	**RMSlog**	$D1_{all}$	<1.25	$<1.25^2$	$<1.25^3$
Supervised	Mayer et al. [43]	—	—	—	—	4.34%	—	—	—
	Wang et al. [45]	—	—	—	—	4.32%	—	—	—
	Zbontar et al. [41]	—	—	—	—	3.89%	—	—	—
	Seki et al. [46]	—	—	—	—	3.09%	—	—	—
	Kendall et al. [47]	—	—	—	—	2.87%	—	—	—
	Luo et al. [43]	—	—	—	—	2.56%	—	—	—
	Pang et al. [48]	—	—	—	—	**2.67%**	—	—	—
Unsupervised	Zhong et al. [50]	—	—	—	—	3.57%	—	—	—
	Godard et al. [49]	0.148	1.344	5.927	0.247	—	0.803	0.922	0.964
	Yang et al. [52]	0.109	1.004	6.232	0.203	—	0.853	0.937	0.975
	Liu et al. [51]	0.051	0.532	3.780	0.126	—	0.957	0.982	0.991
	Wang et al. [53]	**0.049**	**0.515**	**3.404**	**0.121**	**5.943%**	**0.965**	**0.984**	**0.992**
	Jiao et al. [54]	**0.049**	**0.522**	**3.461**	**0.120**	—	**0.961**	**0.984**	**0.992**

Table 4.
Result comparisons of depth evaluation from stereo-based methods on the KITTI dataset. Best performance is marked with bold fonts.

learning optical flow and stereo depth map. By jointly estimating optical flow and stereo depth using unsupervised deep learning like in refs. [52, 53], the network can exploit the shared features and dependencies between the two tasks, leading to improved performance compared to separate estimation methods. Exploiting segmentation in the context of stereo motion learning can lead to further improvements in depth estimation as in ref. [51]. **Table 4** summarizes the evaluation of depth estimation from state-of-the-art stereo-motion-based methods using KITTI [14] dataset.

5.1 Loss function

When learning depth from stereo or multi-view images, we can use a loss function that compares the predicted depth map with the ground truth depth map derived from stereo or multi-view disparity information. One commonly used loss function is the smooth L1 loss, which is defined as:

$$L_{1;smooth} = \begin{cases} |Z(x,y) - Z^{\star}(x,y)| - \frac{\alpha}{2} & if |Z(x,y) - Z^{\star}(x,y)| > \alpha \\ \frac{1}{2\alpha}(Z(x,y) - Z^{\star}(x,y))^2 & otherwise \end{cases} \quad (8)$$

where $Z(x,y)$ is the depth map predicted by the model and $Z^{\star}(x,y)$ is the ground truth depth map derived from stereo or multi-view disparity information. The smooth L1 loss provides a balance between the L1 loss (absolute difference) and the L2 loss (squared difference). It reduces the impact of outliers while still providing gradient information for training. In stereo depth estimation, the ground truth depth map can be obtained by converting the disparity map (derived from stereo correspondence) to depth using the camera parameters and baseline distance. The disparity map represents the horizontal pixel shifts between corresponding points in the stereo images. For multi-view depth estimation, we can use multiple views (more than two) of the same scene to derive the ground truth depth map by triangulation. By estimating the disparity or correspondence between each view and a reference view, we can triangulate the 3D points and obtain the ground truth depth map. Additionally, we can incorporate other regularization terms or constraints into the loss function to further improve the depth estimation. Some common techniques include incorporating geometric consistency, enforcing smoothness or sparsity, or leveraging semantic information. Remember that the choice of the loss function and additional constraints may vary depending on the specific requirements and characteristics of stereo or multi-view depth estimation task.

6. Conclusion

Deep learning methods have significantly advanced the field of depth estimation by providing effective approaches for inferring depth from various types of input data. They are highly data-driven and excel in learning complex patterns and representations from large-scale datasets, enabling them to capture intricate depth cues and generalize well to different scenes and scenarios. Deep learning allows for end-to-end learning, where the model learns to directly predict depth from input data, such as monocular images, stereo pairs, or multi-view images. This eliminates the need for explicitly designing handcrafted features or intermediate steps in the depth estimation pipeline. In the case of monocular case, learning models can estimate depth from a single image, which is a challenging task due to the inherent ambiguity. Despite the limitations, many approaches have achieved impressive results by leveraging large-scale annotated datasets and incorporating various techniques like multi-scale processing, context aggregation, and geometric constraints. Deep learning methods have also shown remarkable success in-depth estimation from stereo and multi-view images. By utilizing the correspondence or disparity information between multiple views, deep models can leverage geometric constraints to provide accurate depth estimation. The choice of loss functions and regularization techniques plays a crucial role in training deep learning models for depth estimation. Common loss functions include mean squared error (MSE) loss, smooth L1 loss, and photometric loss. Regularization techniques like smoothness regularization, geometric consistency, and semantic guidance can further enhance the quality of depth estimation. Pretrained models on large-scale datasets, such as KITTI, have been successfully applied to depth estimation tasks, while transfer learning allows leveraging the knowledge learned from a source task to improve performance on a target depth estimation task with limited data. The choice of deep learning methods for depth estimation depends on the specific application requirements. Factors such as real-time performance, accuracy, robustness to noise and occlusions, and memory efficiency need to be considered when selecting or designing deep learning models for depth estimation. Overall, deep learning methods have revolutionized depth estimation by providing powerful techniques that can learn depth from different input modalities, generalize well to diverse scenes, and achieve state-of-the-art performance. Ongoing research continues to refine and enhance these methods, making depth estimation an active and evolving area of study.

Acknowledgements

This project was supported in part by collaborative research funding from the National Research Council of Canada's Artificial Intelligence for Logistics Program.

Author details

Hicham Sekkati[1]* and Jean-Francois Lapointe[2]

1 National Research Council of Canada, Montreal, Canada

2 National Research Council of Canada, Ottawa, Canada

*Address all correspondence to: hicham.sekkati@nrc-cnrc.gc.ca

References

[1] Calvi GM, Moratti M, O'Reilly GJ, Scattarreggia N, Monteiro R, Malomo D, et al. Once upon a time in Italy: The tale of the Morandi bridge. Structural Engineering International. 2019;**29**(2): 198-217, Taylor Francis

[2] Saxena A, Sun M, Ng AY. Make3D: Learning 3D scene structure from a single still image. IEEE Transactions on Pattern Analysis and Machine Intelligence. 2009;**31**(5):824-840

[3] Zhang R, Tsai PS, Cryer JE, Shah M. Shape-from-shading: A survey. IEEE Transactions on Pattern Analysis and Machine Intelligence. 1999;**21**(8): 690-706

[4] Pentland AP. A new sense for depth of field. IEEE Transactions on Pattern Analysis and Machine Intelligence. July 1987;**9**(4):523-531

[5] Kulkarni JB, Sheelarani CM. Generation of Depth Map Based on Depth from Focus: A Survey, International Conference on Computing Communication Control and Automation - ICCUBEA. Pune, India: IEEE Xplore; 2015. pp. 716-720

[6] Srinivasan P, Garg R, Wadhwa N, Ng R, Barron J. Aperture Supervision for Monocular Depth Estimation, IEEE/CVF Conference on Computer Vision and Pattern Recognition - CVPR. Salt Lake City, UT, USA; 2018. pp. 6393-6401

[7] Suwajanakorn S, Hernandez C, Seitz SM. Depth from Focus with your Mobile Phone, IEEE Conference on Computer Vision and Pattern Recognition - CVPR. Boston, MA, USA; 2015

[8] Hazirbas C, Soyer SG, Staab MC, Leal-Taixé L, Cremers D. Deep depth from focus. In: 14th Asian Conference on Computer Vision - ACCV. Perth, Australia: ACCV; 2018

[9] Gur S, Wolf L. Single Image Depth Estimation Trained Via Depth from Defocus Cues, IEEE/CVF Conference on Computer Vision and Pattern Recognition - CVPR. Long Beach, CA, USA; 2019

[10] Maximov M, Galim K, Leal-Taixe L. Focus on Defocus: Bridging the Synthetic to Real Domain Gap for Depth Estimation, IEEE/CVF Conference on Computer Vision and Pattern Recognition - CVPR. Seattle, WA, USA; 2020

[11] Won C, Jeon HG. Learning Depth from Focus in the Wild, 17th European Conference Computer Vision – ECCV. Tel Aviv, Israel; 2022

[12] Eigen D, Puhrsch C, Fergus R. Depth map prediction from a single image using a multi-scale Deep Network. In: Proceedings of the 27th International Conference on Neural Information Processing Systems - NIPS14. Vol. 2. Montreal, Canada: NIPS14; 2014. pp. 2366-2374

[13] Silberman N, Hoiem D, Kohli P, Fergus R. Indoor Segmentation and Support Inference from RGBD Images. Firenze, Italy: Computer Vision – ECCV; 2012

[14] Geiger A, Lenz P, S. C, Urtasun R. Vision meets robotics: The KITTI dataset. The International Journal of Robotics Research. 2013;**32**(11):1231-1237

[15] Liu F, Shen C, Lin G. Deep Convolutional Neural Fields for Depth Estimation from a Single Image, arXiv, 2014

[16] Saxena A, Chung SH, Andrew YN. 3-D depth reconstruction from a single still image. International Journal of Computer Vision - IJCV. 2008;**76**(1): 53-69

[17] Wang P, Shen X, Lin Z, Cohen S, Price B, Yuille A. Towards unified depth and semantic prediction from a single image. In: IEEE Conference on Computer Vision and Pattern Recognition - CVPR. Boston, MA; 2015

[18] Zoran D, Isola P, Krishnan D, Freeman WT. Learning ordinal relationships for mid-level vision. In: IEEE International Conference on Computer Vision - ICCV. Santiago, Chile; 2015

[19] Chakrabarti A, Shao J, Shakhnarovich G. Depth from a Single Image by Harmonizing Overcomplete Local Network, arXiv-CoRR. 2016

[20] Anirban R, Sinisa T. Monocular depth estimation using neural regression Forest. In: IEEE Conference on Computer Vision and Pattern Recognition - CVPR. Las Vegas; 2016

[21] Laina I, Rupprecht C, Belagiannis V, Tombari F, Navab N. Deeper Depth Prediction with Fully Convolutional Residual Networks, 3DV. California, USA; 2016

[22] Cao Y, Wu Z, Shen C. Estimating Depth from Monocular Images as Classification Using Deep Fully Convolutional Residual Networks, arXiv-CoRR. 2016

[23] Xu D, Ricci E, Ouyang W, Wang X, Sebe N. Multi-scale continuous CRFs as sequential deep networks for monocular depth estimation. In: IEEE Conference on Computer Vision and Pattern Recognition - CVPR. Honolulu, HI, USA; 2017

[24] Anwar S, Hayder Z, Porikli F. Depth estimation and blur removal from a single out-of-focus image. In: British Machine Vision Conference - BMVC. London, UK; 2017

[25] Carvalho M, Le Saux B, Trouve-Peloux P, Almansa A, Champagnat F. Deep Depth from Defocus: How Can Defocus Blur Improve 3D Estimation Using Dense Neural Networks. Munich, Germany: ECCV; 2018

[26] Zhang A, Sun J. Joint depth and defocus estimation from a single image using physical consistency. IEEE Transactions on Image Processing. 2021; **30**:3419-3433

[27] Vijayanarasimhan S, Ricco S, Schmid C, Sukthankar R, Fragkiadaki K. SfM-Net: Learning of Structure and Motion from Video, arXiv-CoRR. 2017

[28] Kumar ACS, Bhandarkar SM, Prasad M. DepthNet: A recurrent neural network architecture for monocular depth prediction. In: IEEE/CVF Conference on Computer Vision and Pattern Recognition Workshops - CVPRW. Salt Lake City, USA; 2018

[29] Fischer P, Dosovitskiy A, Ilg E, Hausser P, Hazırbas C, Golkov V. FlowNet: Learning Optical Flow with Convolutional Networks, arXiv-CoRR, 2015

[30] Garg R, Kumar BG, Carneiro G, Reid I. Unsupervised CNN for Single View Depth Estimation: Geometry to the Rescue, Computer Vision – ECCV. 2016

[31] Zhou T, Brown M, Snavely N, Lowe DG. Unsupervised learning of depth and ego-motion from video. In: IEEE Conference on Computer Vision and Pattern Recognition - CVPR. Honolulu, HI, USA; 2017

[32] Mahjourian R, Wicke M, Angelova A. Unsupervised learning of depth and ego-motion from monocular video using 3d geometric constraints. In: IEEE Conference on Computer Vision and Pattern Recognition - CVPR. Salt Lake City, Utah, USA; 2018

[33] Yin Z, Shi J. Geonet: Unsupervised learning of dense depth, optical flow and camera pose. In: IEEE Conference on Computer Vision and Pattern Recognition - CVPR. Salt Lake City, Utah, USA; 2018

[34] Ilg E, Mayer N, Saikia T, Keuper M, Dosovitskiy A, Brox T. FlowNet 2.0: Evolution of optical flow estimation with deep networks. In: IEEE Conference on Computer Vision and Pattern Recognition - CVPR. Honolulu, HI, USA; 2017

[35] Zou Y, Luo Z, Huang JB. Df-Net: Unsupervised Joint Learning of Depth and Flow Using Cross-Task Consistency, in European Conference on Computer Vision - ECCV. Munich, Germany; 2018

[36] Dosovitskiy A, Fischer P, Ilg E, Hausser P, Hazirbas C, Golkov V, et al. FlowNet. In: Learning Optical Flow with Convolutional Networks, IEEE International Conference on Computer Vision - ICCV. Santiago, Chile; 2015

[37] Casser V, Pirk S, Mahjourian R, Angelova A. Depth prediction without the sensors: Leveraging structure for unsupervised learning from monocular videos. In: Thirty-Third Conference on Artificial Intelligence - AAAI. Honolulu, Hawaii, USA; 2019

[38] Chen Y, Schmid C, Sminchisescu C. Self-supervised learning with geometric constraints in monocular video: Connecting flow, depth, and camera. In: IEEE/CVF International Conference on Computer Vision -ICCV. Seoul, Korea; 2019

[39] Ranjan A, Jampani V, Balles L, Kim K, Sun D, Wulff J, et al. Competitive collaboration: Joint unsupervised learning of depth, camera motion, optical flow and motion segmentation. In: IEEE Conference on Computer Vision and Pattern Recognition -CVPR. Long Beach, California, USA; 2019

[40] Sekkati H, Lapointe J-F. Back to old constraints to jointly supervise learning depth, camera motion and optical flow in a monocular video. In: IEEE International Conference on Image Processing -ICIP. Bordeaux, France; 2022

[41] Žbontar J, LeCun Y. Stereo matching by training a convolutional neural network to compare image patches. arXiv-CoRR. 2016

[42] Sekkati H, Mitiche A. Joint dense 3d interpretation and multiple motion segmentation of temporal image sequences: A variational framework with active curve evolution and level sets. In: International Conference on Image Processing -ICIP. Singapore; 2004

[43] Mayer N, Ilg E, Hausser P, Fischer P, Cremers D, Dosovitskiy A, et al. A large dataset to train convolutional networks for disparity, optical flow, and scene flow estimation. In: IEEE Conference on Computer Vision and Pattern Recognition -CVPR. Las Vegas, Nevada, USA; 2016

[44] Luo W, Schwing AG, Urtasun R. Efficient deep learning for stereo matching. In: IEEE Conference on Computer Vision and Pattern Recognition -CVPR. Las Vegas, NV, USA; 2016

[45] Wang Z, Zhu S, Li Y, Cui Z. Convolutional neural network based deep conditional random fields for stereo matching. Journal of Visual

Communication and Image Representation. 2016;**40**:739-750

[46] Seki A, Pollefeys M. SGM-nets: Semi-global matching with neural networks. In: IEEE Conference on Computer Vision and Pattern Recognition -CVPR. Honolulu, HI, USA; 2017

[47] Kendall A et al. End-to-end learning of geometry and context for deep stereo regression. In: IEEE International Conference on Computer Vision -ICCV. Venice, Italy; 2017

[48] Pang J, Sun W, Ren JSJ, Yang C, Yan Q. Cascade residual learning: A two-stage convolutional neural network for stereo matching. In: IEEE International Conference on Computer Vision Workshop -ICCVW. Seoul, Korea; 2019

[49] Godard C, Mac Aodha O, Brostow GJ. Unsupervised monocular depth estimation with left-right consistency. In: IEEE Conference on Computer Vision and Pattern Recognition -CVPR. Honolulu, Hawaii, USA; 2017

[50] Zhong Y, Dai Y, Li H. Self-supervised learning for stereo matching with self-improving ability. In: IEEE Conference on Computer Vision and Pattern Recognition -CVPR. Honolulu, Hawaii, USA; 2017

[51] Liu L, Zhai G, Ye W, Liu Y. Unsupervised Learning of Scene Flow Estimation Fusing with Local Rigidity, 28th International Joint Conference on Artificial Intelligence -IJCAI. Macao, China; 2019

[52] Yang Z, Wang P, Wang Y, Xu W, Nevatia R. Every Pixel Counts: Unsupervised Geometry Learning with Holistic 3d Motion Understanding, in European Conference on Computer Vision -ECCVW. Germany: Munich; 2018

[53] Wang Y, Wang P, Yang Z, Yi CL. Unos: Unified unsupervised optical-flow and stereo-depth estimation by watching videos. In: IEEE Conference on Computer Vision and Pattern Recognition -CVPR. Long Beach, California; 2019

[54] Jiao Y, Tran T, Shi G. EffiScene: Efficient per-pixel rigidity inference for unsupervised joint learning of optical flow, depth. In: Camera Pose and Motion Segmentation, IEEE/CVF Conference on Computer Vision and Pattern Recognition -CVPR. Nashville, TN, USA; 2021

Chapter 11

Immune Neural Network Machine Learning of Autonomous Drones for Energy Efficiency and Collision Prevention

Mikhail Gorobetz, Leonids Ribickis, Anna Beinarovica and Aleksandrs Korncjevs

Abstract

The chapter is related to the safe and energy-efficient motion of autonomous drones and describes the developed novel immune neural network-based machine learning technology for its control. The technology is inspired by two biological systems – immune system and neural networks and their artificial analogs and evolutionary theory. The developed novel mathematical models and algorithms for this technology allow skipping the preliminary supervised training step and adapted for real-time continuous unsupervised self-learning of the drone to recognize the dangerous situation, to prevent the collision by making control decisions autonomously and continuously learning to keep optimal energy consumption during the motion. The chapter includes the study of existing neural network-based solutions for the recognition and prevention of dangerous situations and energy efficiency of drones, describes the developed target function and algorithm for immune neural network machine learning technology and simulation and experimental results proving the efficiency of this technology.

Keywords: machine learning, neural networks, energy efficiency, safety, autonomous vehicles, collision prevention, distributed system, immune memory

1. Introduction

Drones have already been integrated into numerous activities, including inspecting hazardous objects, delivery services, people searching, fulfilling military objectives, agriculture, etc. Furthermore, photographers worldwide are increasingly embracing modern methods and technologies to enhance their photography during celebratory events. An interesting highlight is that in Australia, drones come equipped with sophisticated artificial intelligence-driven software, enabling them to distinguish boats and different marine life in real-time with an impressive 90% accuracy [1]. Amazon legally delivered its first Prime order in the United Kingdom in December 2016 [2]. And the use of drones still continues to spread to various areas.

Although most drones are currently controlled by human operators using remote controls to dictate their speed and trajectory, this approach may not be the most productive use of human time and effort in an era focused on automation and optimization. Additionally, relying on human operators leaves room for potential mistakes, which could lead to undesirable consequences. That is why the use of intelligence systems in drones is developing also.

For instance, the Patroller multi-sensor long endurance quadcopter was specifically designed to cater to the requirements of armed forces, intelligence operations, operational support, and maritime surveillance. What sets it apart is its autonomous launch system, enabling it to take off from airports without any need for adjustments to ground facilities. With an impressive endurance of 20 hours and a payload capacity of up to 250 kg, the Patroller is well-suited for prolonged missions [3].

There are bunches of areas where drones are used: military purposes; patrolling and searching tasks; order delivery; mapping of landslides, hurricanes or other cataclysms affected area; infested crop damage assessment; 3-dimensioinal terrain model construction, etc.

Such tasks as patrolling or searching require a team of drones working in the same area. When dealing with tasks that involve multiple drones, there is an added risk of potential collisions between them. For example, if multiple drones are assigned to search for the same object, and each drone is programmed to reach its own specific destination, there is a possibility that all drones may converge on nearly the same coordinates, increasing the risk of a collision. Another problem with the use of autonomous drones is untimely battery discharge, which can lead to the loss of the drones. That is why the problem of the safe and energy-efficient drive of autonomous drones is important and needs to be well studied.

Analysis of the research base reveals that currently, the majority of energy-saving solutions are related to route calculation and trajectory planning for unmanned vehicles, as well as energy-saving algorithms for other equipment unrelated to electric drive systems. Only a few studies focused on reducing energy consumption explore the mechanical and electrical properties of power devices, but not connected with autonomous drones and machine learning.

The scientific novelty of this research is the technology for the drone embedded system, which is able to learn in real-time to avoid crashes and control the motion with optimal energy consumption without any preliminary datasets, pretraining, and a teacher.

Authors named it as immune neural network (INN). It is a symbiosis that uses the best features of artificial neural networks (ANN) and artificial immune system (AIS). Separately each of them has a bunch of disadvantages. ANN requires preliminary training, may be overtrained, etc. AIS does not require training, but the optimization requires time and may not be used for real-time control. Novel INN system combines the possibilities of ANN and AIS is adapted for real-time safety and energy control.

The main hypothesis is that an immune neural network can make control decisions to prevent drone collisions with better performance than a traditional neural network in this task and is able to reduce the energy consumption of autonomous unmanned vehicles – drones.

The first part of the chapter contains an analysis of the existing types of solutions of the problem. The second part is devoted to the developing the novel system structure, which could help to make the proposed system cheaper, faster, and easier to implement. The target function of the proposed novel INN is described in the third part of the chapter. The fourth part is devoted to the proposed algorithm, based on the novel INN machine learning technology. The last part of the chapter is devoted to the results of testing the efficiency of the developed algorithm. The developed algorithm was tested with the help of computer modeling.

2. Review of the scientific research and literature about autonomous electric transport control

A research paper [4] previously proposed a method for real-time traffic flow parameter estimation from unmanned aerial vehicle (UAV) video, which relied on an ensemble classifier and optical flow. In contrast, the system presented in this study eliminates the need for additional infrastructure embedded objects, or/and devices. Two other articles [5, 6] also delve into motion control tasks for UAVs, both referencing the Lyapunov method. Comparing the motion control of aerial vehicles with surface transport allows concluding that of surface autonomous vehicles have a higher probability of collision because there is no ability to change the altitude component. A hierarchical collision-avoidance strategy is proposed as a theoretically approach for high-speed self-driving vehicle collision-avoiding task solving [7]. Fuzzy logic and FMCW radar (Frequency-Modulated Continuous Wave) usability for motion control task solving are described in research [8], but this system also is not provided with surrounding environment visual monitoring elements or sensors. The author presented a doctoral thesis in ref. [9], that is devoted to knowledge extraction from trained ANN. Research described in this research proposes a novel ANN, that does not require previous training, the proposed system is unsupervised and does not need a teacher.

In the research discussed in ref. [10], novel drone-following models are introduced for effective drone management within the transportation management system. These models operate on the fundamental premise that drones navigate by following a leading drone. But in difference with this research, in ref. [10] several important situations were not considered, like safe distance is measured not only in the drone directly in front but also in two drones beside, or increasing the number of drones. Intelligent electric vehicle motion and crossroad control using genetic algorithms is proposed in paper [11].

The book published, that was devoted to the scientific results: "Unmanned Electrical Vehicles and Autonomous System Simulation". This book and its offshoots were prepared to provide a comprehensive introduction to the domain of the autonomous system [12]. The book was also mentioned artificial intelligence for autonomous drives, but no specific solutions, like ANN, were mentioned there.

The conventional ANN requires adjustments or fine-tuning to adapt to individual situations or tasks. In simpler terms, the traditional ANN uses the current weights as a starting point for the next round of re-training. However, this process can be time-consuming due to the wide variety of situations it needs to handle [13]. The INN proposed in this study includes an immune memory (IM) that retains information from past successfully prevented collision scenarios. By referencing this memory to find similar situations and utilizing the stored weights as a foundation for re-training, the INN can considerably minimize the time needed for self-tuning when dealing with a similar problem. In ref. [14] a similar solution is proposed. The main idea of that research is a retraining of a static neural network, by using IM-based algorithm. In the solution presented in ref. [14], the approach involves creating the initial population (a set of chromosomes) by random means. This means that the population consists of fuzzy weights for ANN, and these weights are assigned random values during the generation process. In difference with that research, the main idea of this research is that input data are stored in the IM together with weights, that was used previously for solving this situation, that helps to reduce calculation time, which is very important for real-time systems, thus to reduce energy consumption. Also, the proposed novel algorithm allows retraining the neural network. In this research, no initial population is needed. Training can start even with an empty IM. The found solutions will be those that will be written in the IM and further adjusted.

3. System structure

3.1 General safety system structure for drone dangerous situation recognition and prevention task

System models can be divided into centralized, decentralized, and distributed.

In a centralized system, all computing power may be allocated to one user when no other users are attached to the system. Consequently, the execution time of all users' applications will be increased if the mainframe serves many users [15].

In simpler terms, decentralization is a system model where decision-making occurs at different levels. In this approach, the larger system is subdivided into smaller segments or groups, allowing for a more manageable assessment of the system's performance and the contributions of individuals within each group [16].

A distributed operating system is a unique type of operating system that spans across multiple objects. The goal is to offer a useful range of services, making all the machines appear and function as a unified single entity. These distributed operating systems collaborate on all the machines they manage, which can either operate independently or serve as resources within the broader distributed system [17].

Centralized, decentralized, and distributed system models were compared. The results of the comparison show that a distributed system is preferable because it is easier to implement and has less components, it is cheaper for infrastructure owners and it is not connected to the specific area. It has also a decreased time for reaction and decreased risk of system failure. That is why a distributed system was chosen for this research.

Figure 1 illustrates the schematic representation of the distributed system's structure. Where: SAT – satellite, is a component that serves to acquire the real-time coordinates of an object's position; Drone – autonomous drone; $Drone_n$ – other autonomous drones; MECH – mechanical part that is responsible for the drones moving; IO – infrastructure objects; CO – control system; SEN – sensor; EC – proposed novel immune neural network energy efficiency control module; SC – proposed novel immune neural network safety module; ORM – object recognition module, is using incoming video data for CNN to recognize other drones and drones that do not have the proposed embedded device as well as other objects in the visibility range; LC – different type of level crossings, if necessary; RS – signs, that limit the possible movement parameters on the concrete road/way section, if necessary; TL – different types of the traffic lights, if necessary; GNSS – GNSS receiver, is utilized to receive and determine the precise coordinates of the drone's current location; RFM – serves

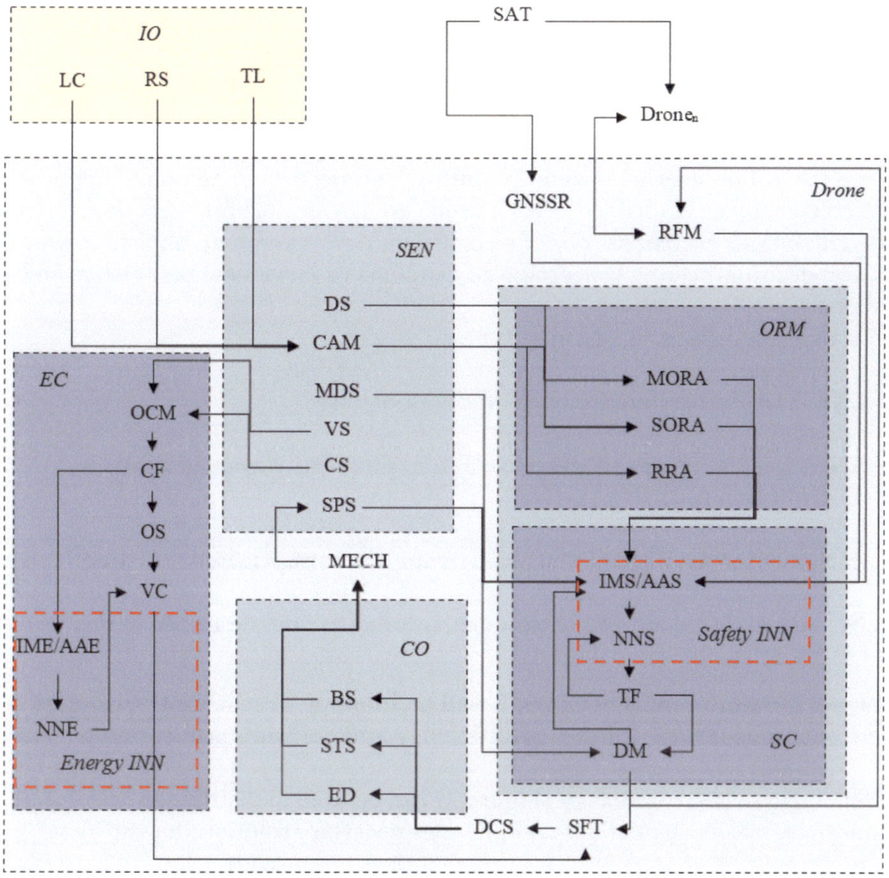

Figure 1.
Scheme of the general safety system structure for drone dangerous situation recognition and prevention task.

the purpose of communication with other devices embedded in different drones and infrastructure objects; DS – distance sensor; CAM – camera is used to capture images of the surrounding environment; MDS – movement direction sensor, to receive the data of the drone movement direction; VS – voltage sensor; CS – current sensor; SPS – speed sensor to receive the data of the drone speed and acceleration; MORA – moving object recognition algorithm, to recognize moving objects such as human being, animals, other drones in the visibility range, based on the convolutional neural network (CNN); SORA – static object recognition algorithm, to recognize static objects such as fallen tree or broken car on the way of motion, based on the CNN; RRA – road recognition algorithm, to recognize the presence of turns and road profile, based on the CNN; IMS – immune memory for safety control; AAS – affinity algorithm for safety control; NNS – neural network as a part of the safety INN; DM – decision making module is responsible for making decisions to minimize risks and control the drone's movement. It calculates appropriate curves and actions to achieve the desired speed and reach the target point on the designated route; TF – target function; CF – constraint formation to control signals; OS –optimal search algorithm module; VC – validation of the control signals; IME – immune memory for energy control; AAE – affinity algorithm for energy control; NNE – neural network as a part of Energy INN; SFT – safety check to avoid conflicts between safety control and energy control, where safety is a priority; DCS – drone control system, to control actions of the driver. If the driver receives the indications on the information display but does not react in time, then the driver control system sends the signal to the drone control system, and the proposed system intervenes in the drone control; BS – drone braking system, to control the stop of the drone, if applicable; ED – drone electric drive; STS – drone steering system to control the trajectory of motion, if applicable.

Microcontrollers embedded in each drone are responsible for executing various tasks. These tasks encompass object recognition, risk assessment, opportunity evaluation, and decision-making concerning adjustments to movement parameters and control signals.

The suggested system's structure is based on the:

- CNN-based subsystem for object recognition task;
- ANN-based subsystem for drone collision probability evaluation and minimization task;
- INN-based technology of machine learning for unsupervised safe drone control.

CNN and ANN are already known methods and are not described in this study.

3.2 Novel immune neural network-based technology of machine learning of autonomous drones for energy efficiency and collision prevention

Novel INN-based technology (**Figure 2**) can be used in distributed systems. It obtains data, makes calculations and provides necessary solutions how to avoid collision in the context of one particular drone. It does not provide solutions for other participants.

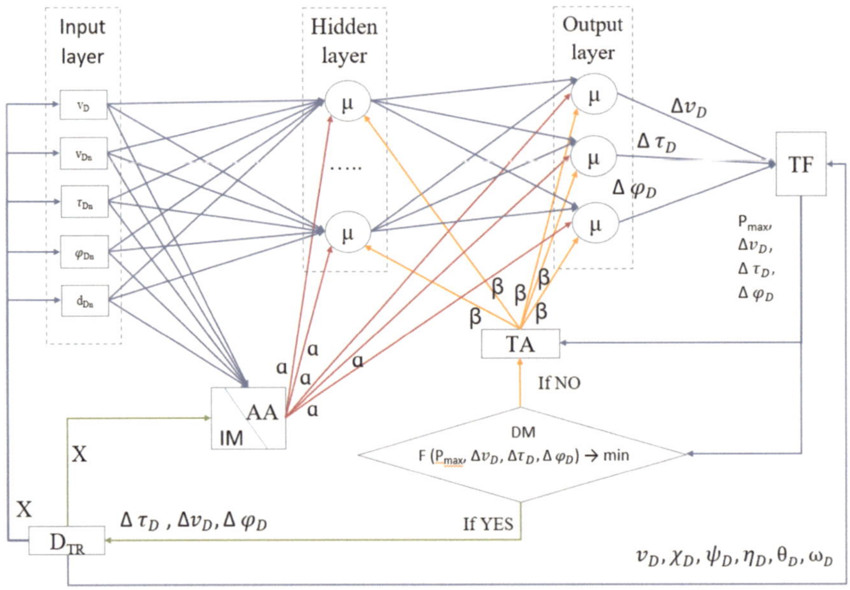

Figure 2.
Immune neural network system structure of the drone.

3.2.1 Input data (X)

Data:

- v_D – speed of the own drone;
- v_{Dn} – speed of other drones;
- τ_{Dn} – horizontal movement direction of the other drone relative to one's own direction;
- φ_{Dn} – vertical movement direction of the other drone relative to one's own direction;
- d_{Dn} – distance till the possible crossing point with the other drone.

Number of parameters in the input data (X) depends on the situation – the number of other drones in the control area of own drone. There is one input parameter for own drone: speed. There are four input parameters for other drones: speed, horizontal movement direction in relation to the own drone, vertical movement direction in relation to the own drone, and distance till possible crossing point.

Data is received from the drone-embedded electronic device and is sent to the input layer of the immune neural network INN.

3.2.2 Input layer

The input layer of the INN receives input data (X).

Each drone takes into account only those drones, which are in its control area, in case to minimize the number of necessary calculations.

Input data (X) is ordered for a more accurate recognition of the situation. The goal is to order multiple drones in relation to the own drone to better understand their position and relative movement. Three parameters are used for ordering the drones: the horizontal movement direction τ_{Dn}, the vertical movement direction φ_{Dn}, and the distance to the crossing point d_{Dn}. Ordering of other drones is done according to the slope τ_{Dn} to these objects, starting from 0°, clockwise. If multiple drones have the same value of τ_{Dn}, then the ordering of these drones is done according to the slope φ_{Dn} to these objects, starting from 0°, clockwise. If multiple drones have the same value of φ_{Dn}, then the ordering of these drones is done according to the distance to the crossing point with these drone d_{Dn}. This method helps to describe the situation accurately.

Input data (X) are sent from the input layer to the affinity algorithm AA and hidden layer.

3.2.3 Affinity algorithm (AA)

The affinity algorithm AA examines all analogous situations stored in the immune memory IM and computes their errors (discrepancies) $\mathcal{E}$. Situation with a smallest error (discrepancy) $\mathcal{E}$ is chosen and its identification number α is distributed to all μ neurons of the proposed INN. If there is no similar situation stored in IM immune memory, then situation number $\alpha = 0$ is sent to the μ neurons.

3.2.4 Immune memory (IM)

Database, that contains input data about previous situations that were solved. Each situation has its' number α.

To enhance the efficiency of the affinity algorithm (AA) and the storage of data in the immune memory (IM), a clustering technique is implemented. This approach avoids unnecessary searches for similar situations in scenarios with different numbers of participants. This ensures that, for instance, if four drones are involved in a potential collision scenario, there is no need to search for a similar situation among cases with only three participants. Clustering allows for faster and more efficient matching processes in the system.

3.2.5 Hidden layer

The hidden layer consists of specialized μ neurons.

Input data of each μ neuron of the hidden layer is:

- input data (X);
- situation number α received from the affinity algorithm AA;
- signal β that indicates the need to recalculate the weights of μ neurons and is received from the training algorithm TA.

In the μ neuron number of situation α is stored together with a set of weights W_μ, that were used while solving the exact problem i.e. processing the similar input data. After number of the situation α is received, weights W_μ are chosen and training can be started. If there is no similar situation and $\alpha = 0$, then $W_\mu = 0$.

3.2.6 Output layer

Output layer consists of specialized μ neurons.

Input data of each μ neuron of the output layer is:

- output data of the μ neurons of the hidden layer;
- situation number α received from the affinity algorithm AA;
- signal β that indicates the need to recalculate the weights of μ neurons and is received from the training algorithm TA.

In the μ neuron number of situation α is stored together with set of weights W_μ, that were used while solving the exact problem, similar as in the μ neuron of the hidden layer. After number of the situation α is received, weights W_μ are chosen and training can be started. If there is no similar situation and $\alpha = 0$, then $W_\mu = 0$.

Output data of the output layer:

- necessary horizontal movement direction change of the own drone $\Delta\tau_D$;

- necessary vertical movement direction change of the own drone $\Delta\varphi_D$;
- necessary speed change of the own drone Δv_D.

3.2.7 Target function (TF)

Input data of target function TF:

- necessary horizontal movement direction change of the own drone $\Delta\tau_D$;
- necessary vertical movement direction change of the own drone $\Delta\varphi_D$;
- necessary speed change of the own drone Δv_D;
- input data obtained directly from drone-embedded electronic device:
    - v_D – actual speed of the drone;
    - χ_D – latitude of the drone's actual position;
    - ψ_D – longitude of the drone's actual position;
    - η_D – altitude of the drone's actual position;
    - θ_D – actual horizontal movement direction of the drone;
    - ω_D – actual vertical movement direction of the drone.

In the present study, the crossing points' locations are variable, which adds complexity to the solution. The identified solution $< \Delta v_D, \Delta\tau_D, \Delta\varphi_D >$ is affected by the distance to the crossing point. Thus, the evaluation of the target function TF requires the additional inputs $< v_D, \chi_D, \psi_D, \eta_D, \theta_D, \omega_D >$ obtained directly from the drone to re-calculate the crossing point, distance, and time to it.

Target function TF calculates the collision probability P_{max}.

Output data of target function TF:

- collision probability P_{max};
- necessary horizontal movement direction change $\Delta\tau_D$;
- necessary vertical movement direction change $\Delta\varphi_D$;
- necessary speed change Δv_D.

3.2.8 Decision module (DM)

Input data of decision module DM:

- collision probability P_{max}, received from the target function TF;

- necessary horizontal movement direction change $\Delta\tau_D$, received from the target function TF;
- necessary vertical movement direction change $\Delta\varphi_D$, received from the target function TF;
- necessary speed change Δv_D, received from the target function TF.

Decision module (DM) evaluates the found solution.

If collision probability P_{max} is greater than acceptable (safe) collision probability P_{safe} and

- If number of training iterations t is less than maximal possible number of iterations T_{max}, that means the solution is not found yet and training must be repeated. Decision making module (DM) sends signal to the training algorithm (TA).
- If number of training iterations t is bigger or equal than maximal possible number of iterations T_{max}, that means that the situation can not be solved in the defined time, so speed reduction is done. Decision-making module (DM) sends signal to the drone-embedded electronic device to stop the drone $v_D = 0$.

If collision probability P_{max} is less or equal than acceptable (safe) collision probability P_{safe}, then found solution $< \Delta v_D, \Delta\tau_D, \Delta\varphi_D >$ is sent to the drone-embedded electronic device and match error ε_a between the current situation and situation chosen from the immune memory IM in the beginning of training is calculated:

- if match error ε_a is bigger than maximal possible match error ε_{lim}, responsible for creation new record in the immune memory IM or replacing the existing, then IM saves the situation as a new record. Each μ neuron of the hidden and output layers saves set of weights W_μ that was used for solving this situation together with a number of this situation α.
- if match error ε_a is less or equal than maximal possible match error ε_{lim}, then the record of the situation α is updated in the immune memory IM. Values of weights W_μ of μ neurons of the hidden and output layers are updated according to the last used.

3.2.9 Training algorithm (TA)

Input data of the training algorithm (TA):

- collision probability P_{max}, received from the target function TF;
- necessary horizontal movement direction change $\Delta\tau_D$, received from the target function TF;
- necessary vertical movement direction change $\Delta\varphi_D$, received from the target function TF;

- necessary speed change Δv_D, received from the target function TF;
- signal to repeat training, received from the decision-making module DM.

The training algorithm is used instead of the traditional backpropagation algorithm. Backpropagation is typically used in supervised learning where the network is trained using labeled data, but the proposed novel INN is based on unsupervised learning.

Training algorithm TA stores the last value of the P_{max}, which was received while solving this situation and compares this value to the new one. TA sends a signal β to all μ neurons, that means that training must be repeated. Signal β differs according to the result of the P_{max} comparison.

If it is the first training iteration, TA does not have information about the previous P_{max}, so TA sends a signal β_1 to all μ neurons of the hidden and output layer. Signal β_1 means that the found solution does not solve the situation and training must continue.

The same happens if the result of the found solution is better or equal than previous $P_{max\,2} \leq P_{max\,1}$. TA sends a signal β_1 to all μ neurons of the hidden and output layer, which means the found solution is not worse than previous one and training must continue.

If the result of the found solution is worse than the previous $P_{max\,2} > P_{max\,1}$, then TA sends a signal β_2 to all μ neurons of the hidden and output layer. Signal β_2 means that the found solution does not solve the situation and the result of the last iteration is worse than the result of the previous one. The values of the weights must be returned to the previous before training continues.

3.2.10 Training of μ neurons

When receiving β_1 new values of weights $W_{\mu j}$ are randomly chosen from the range $\left(W_{\mu j} - z \leq W_{\mu j+1} \leq W_{\mu j} + z\right)$, where z is a predefined range parameter (may be adjustable).

When receiving β_2 *new* values of weights $W_{\mu j}$ rollback to the previous values $W_{\mu j-1}$ and then are randomly chosen from the range $\left(W_{\mu j-1} - z \leq W_{\mu j} \leq W_{\mu j-1} + z\right)$, where z is predefined range parameter (may be adjustable).

4. Target function for the immune neural network

The general target function with anti-collision criteria is shown in Eq. (1), where:

P_{max} - maximal collision probability from the set of probabilities of collision for all pairs of drones;

$\Delta\Theta = (\Delta\theta_1, \ldots, \Delta\theta_n)$, - set direction changes in the horizontal plane of all drones;

$\Delta\Theta = (\Delta\theta_1, \ldots, \Delta\theta_n)$, - set direction changes in the vertical plane of all drones;

$\Delta\Phi = (\Delta\varphi_1, \ldots, \Delta\varphi_n)$ - set of speed changes of all drones;

DIST – safety criterion (Eq. (2)), where: $DIST_{safe}$ is safety distance limit for each pair of drones $<UAV_i, UAV_j>, i = 1..n, j = 1..n, i \neq j$;

P_{IJ} = $(P(<D_1, D_2>), \ldots, P(<D_i, D_j>), \ldots, P(<D_{n-1}, D_n>))$ - set of probabilities of collision for all pairs of drones $<D_i, D_j>$, $i \neq j$, $i,j = 1..n$.

Permissible changes of direction depend on the drone specifications and other circumstances. Restrictions for the horizontal movement direction change $\Delta\tau_i$, restrictions for the movement direction (in the vertical plane) change $\Delta\varphi_i$ and restrictions for the speed change Δv_i were also defined.

$$\begin{cases} P_{max}\left(\chi_c^D, \psi_c^D, \eta_c^D, \Delta\tau, \Delta\varphi, \Delta v\right) = max\left(P_{IJ}\right) \rightarrow min \\ \Delta\tau_\Sigma(\Delta\tau) = \sum_{i=1}^{n} \Delta\tau_i \rightarrow min \\ \Delta\varphi_\Sigma(\Delta\varphi) = \sum_{i=1}^{n} \Delta\varphi_i \rightarrow min \\ \Delta v_\Sigma(\Delta v) = \sum_{i=1}^{n} \Delta v_i \rightarrow min \\ DIST = \left|D_i D_j\right| > S \\ \Delta\tau_1 < \Delta\tau_i < \Delta\tau_2 \\ \Delta\varphi_1 < \Delta\varphi_i < \Delta\varphi_2 \\ \Delta v_1 < \Delta v_i < \Delta v_2 \\ i = 1..n, j = 1..n, i \neq j \end{cases} \quad (1)$$

$$DIST = \left|D_i D_j\right| = \sqrt{\left(\chi_c^j - \chi_c^i\right)^2 + \left(\psi_c^j - \psi_c^i\right)^2 + \left(\eta_c^j - \eta_c^i\right)^2} > DIST_{safe} \quad (2)$$

In case of the distributed system, each i-th drone is looking for its own direction and/or speed change solution $< \Delta\theta_i, \Delta\varphi_i >$, according to the task. Additionally the drone's energy control system search for optimal control signals c_1, c_2, c_3, c_4 to minimize energy consumption E_v and reach the destination point x_M, y_M, z_M.

Thus, the target function for a single drone with energy control is the following:

$$\begin{cases} P_{max}\left(\chi_c^D, \psi_c^D, \eta_c^D, \Delta\tau_0, \Delta\varphi_0, \Delta v\right) = \max\left(P_{0j}\right) \rightarrow min \\ E_v = \int I \cdot U \cdot dt = f\left(c_1, c_2, c_3, c_4, t\right) \rightarrow min \\ \Delta\tau_0 \rightarrow min \\ \Delta\varphi_0 \rightarrow min \\ \Delta v_0 \rightarrow min \\ DIST = \left|D_0 D_j\right| > S \\ \Delta\tau_1 < \Delta\tau_0 < \Delta\tau_2 \\ \Delta\varphi_1 < \Delta\varphi_0 < \Delta\varphi_2 \\ \Delta v_1 < \Delta v_0 < \Delta v_2 \\ j = 1..n \\ \left|x_M - \chi_c^0\right| \leq \varepsilon_X \\ \left|y_M - \psi_c^0\right| \leq \varepsilon_Y \\ \left|z_M - \eta_c^0\right| \leq \varepsilon_Z \\ c_{1min} \leq c_1 \leq c_{1max} \\ c_{2min} \leq c_2 \leq c_{2max} \\ c_{3min} \leq c_3 \leq c_{3max} \\ c_{4min} \leq c_4 \leq c_{4max} \end{cases} \quad (3)$$

where:

P_{max} - represents the highest probability of collision between the own drone D_0 and all other drones within the control area;

$\Delta\Theta = (\Delta\theta_1, \ldots, \Delta\theta_n)$, - direction change in the horizontal plane of the own drone D_0;

$\Delta\Theta = (\Delta\theta_1, \ldots, \Delta\theta_n)$, - direction change in the vertical plane of the own drone D_0;

$\Delta\Phi = (\Delta\varphi_1, \ldots, \Delta\varphi_n)$ - speed change of the own drone D_0;

c_1, c_2, c_3, c_4 – control signals roll, pitch, throttle and yaw;

c_{imin}, c_{imax} - limits of i-th control signal;

$\varepsilon_X, \varepsilon_Y, \varepsilon_Z$ – acceptable precision of coordinates assumed the destination is reached.

DIST - safety criterion (Eq. (4)), where: $DIST_{safe}$ is the safety distance limit for own drone with all the other drones $< UAV_i, UAV_j >, i = 1..n, j = 1..n, i \neq j$;

$P_{0j} = (P(<D_0, D_1>), \ldots, P(<D_0, D_j>), \ldots, P(<D_0, D_n>))$ - set of probabilities of collision between the own drone D_0 and all other drones within the control area, j = 1..n;

Restrictions of the own parameters change $(\Delta\tau_0; \Delta\varphi_0; \Delta v_0)$ were also defined.

$$DIST = |D_0D_j| = \sqrt{\left(\chi_c^j - \chi_c^0\right)^2 + \left(\psi_c^j - \psi_c^0\right)^2 + \left(\eta_c^j - \eta_c^0\right)^2} > DIST_{safe} \tag{4}$$

Function of the decision-making module F_{DM} is represented as follows:

$$F_{DM} = TF(P_{max}, \Delta v_D, \Delta\tau_D, \Delta\varphi_D) \rightarrow min \tag{5}$$

Thus, we can evaluate the result of training the INN without a teacher with the help of the proposed target function and make a decision about accepting the solution or continue training.

5. Algorithm for immune neural network of autonomous drones for collision prevention task

The algorithm of INN consists of the following steps:

STEP 1. Receive input data DAT from n drones located in the area of visibility. These data are locations $<\chi_c^D, \psi_c^D, \eta_c^D>$, speed v^D, horizontal movement direction θ^D, and vertical movement direction ω^D of drones.

STEP 2. The proposed INN requires data about other drones' location in relation to the own drone location. Data DAT needs to be proceeded before it will enter the input layer of the proposed INN.

STEP 2.1. Input data DAT contains coordinates of other drones position $<\chi_c^D, \psi_c^D, \eta_c^D>$, own position is known also. Distances to the possible crossing points with other drones $DIST_D$ are calculated.

STEP 2.2. The next step is to organize the drones for a more precise definition of the situation. The first drone is always drone itself. The other drones are ordered according to their horizontal movement direction τ_n^D in relation to the own drone, starting from 0 degrees and proceeding clockwise. If multiple drones have the same value of τ_n^D, they are then ordered according to their vertical movement direction φ_n^D in relation to the own drone, starting from 0 degrees and proceeding clockwise. If multiple drones have the same value of φ_n^D, they are then ordered according to the distance to the crossing point with the own drone $DIST_n^D$.

Horizontal movement directions of other drones τ_n in relation to the own drone direction and vertical movement directions of other drones φ_n in relation to the own drone direction are calculated in this step. Actions, provided in STEP 2, transform input data DAT to the input data X to be processed by the INN (Eq. 6).

$$X = (v_0, v_1, \tau_1, \varphi_1, d_1 \dots, v_n, \tau_n, \varphi_n, d_n) = (x_0, x_1, x_2, x_3, x_4, \dots, x_{4n-3}, x_{4n-2}, x_{4n-1}, x_{4n}) \quad (6)$$

Where,

n – number of other drones, $n = 0$ – own drone, $n > 0$ – all others drones;

v_n, x_{4n-3} – speed of n-th drone;

τ_n, x_{4n-2} – horizontal movement direction of the n-th drone. Direction of movement of the another drone ($n > 0$) is relative to one's own ($n = 0$) direction, but $\tau_0 = 0$;

φ_n, x_{4n-1}– vertical movement direction of the n-th drone. Direction of movement of the another drone ($n > 0$) is relative to one's own ($n = 0$) direction, but $\varphi_0 = 0$;

d_n, x_{4n} – distance till the possible crossing point of own drone with another drone's $n > 0$ trajectory. Thus $d_0 = 0$.

STEP 3. The calculation of collision probability P is intended to determine whether it is necessary to minimize the risk of collision. If no, end of the algorithm. If yes, then go to the next step.

STEP 4. After input data X enters the input layer of INN, data X is sent to the specialized μ neurons and affinity algorithm (AA). The AA (X, S) checks all situations S stored in the IM $S = \{s_1, s_2, \dots, s_m\}$, calculates the set of discrepancies $\mathcal{E} = (\varepsilon_1, \dots, \varepsilon_k)$, where

$$\varepsilon_j = \sum_{i=0}^{n} \sum_{k=1}^{2} \left(\frac{X_{ik} - X_{ik}^{j}}{X_{ik}} \right)^2 \quad (7)$$

and finds the closest match ε_α, where $\varepsilon_\alpha = \min(\varepsilon)$.

STEP 5. When μ neuron receives input data X, it activates and increases iteration counter $t = t + 1$. When situation number α is received, set of weights W_μ are selected from the memory of the μ neuron. If there is no similar situation in the IM and $\alpha = 0$, then $W_\mu = 0$.

STEP 6. Input data X, situation number α, received from the affinity algorithm AA, and signal β, which indicates the need to recalculate the weights of μ neurons, are the input data of each μ neuron of the hidden layer μ_{IIID}. Feed forward input through the NN is done. Outputs for own vertical movement direction change $O_{\mu p3} = \Delta\varphi_D$, own horizontal movement direction change $O_{\mu p2} = \Delta\tau_D$ and own speed change $O_{\mu p1} = \Delta v_D$ are generated as a result.

STEP 7. The TF calculates the collision probability P_{max}, that is maximal collision probability from the set of probabilities of collision for all pairs of drones P_{IJ}. TF uses updated data, received directly from the drone embedded device D_{TR}.

STEP 7.1. The TF function defines the directions τ^D and φ^D of each drone in relation to the own drone.

STEP 7.2. Next step is to detect the crossing point (χ_p, ψ_p, η_p) in 3D space.

STEP 7.3. Involves checking if the crossing point (χ_p, ψ_p, η_p) is discovered and lies along the path of motion. If it is, proceed to STEP 7.4. Otherwise, proceed to STEP 7.6.

STEP 7.4. To find the distance between the altitudes of the i-th drone and its own drone at the point (χ_p, ψ_p, η_p), the calculation involves $\Delta\eta = \eta_p^i - \eta_p^{own}$.

STEP 7.5. If $\Delta\eta \leq DIST_{safe}$ then it is assumed that a potentially dangerous point exists, and the probability of collision P is calculated.

STEP 7.6. If the crossing point (χ_p, ψ_p, η_p) is not found, then the trajectories are parallel and $DIST_{safe}$ should be checked for safe passing.

STEP 8. If $P_{max} > P_{safe}$, where P_{safe} is the maximal acceptable (safe) collision probability, then is checked either the solution is better than the previous or worse:

If $t = 1$, then signal β is sent to all μ neurons and repeats from STEP 6.

If $1 < t < T_{max}$ and $P_{max\,2} > P_{max\,1}$, then signal β is sent to all μ neurons. μ neurons return the previous values of W_μ and repeat from STEP 6.

If $1 < t < T_{max}$ and $P_{max\,2} \leq P_{max\,1}$, then signal β is sent to all μ neurons and repeat from STEP 6.

If $t \geq T_{max}$, the situation cannot be resolved within the specified timeframe, requiring a safe solution. In this research, the safe solution involves reducing the speed to $\Delta v_i = v$ and implementing the END algorithm. Otherwise, continue to STEP 9.

STEP 9. If $P_{max} \leq P_{safe}$ then:

The calculated speed Δv_D, horizontal and vertical movement directions $\Delta \tau_D$ and $\Delta \varphi_D$ changes are accepted as the solution and sent to the embedded electronic device for the drone control.

Match error ε_a is compared with a maximal possible match error ε_{lim}, responsible for creation new record in the immune memory IM or replacing the existing:

if $\varepsilon_\alpha > \varepsilon_{lim}$, then each μ neuron saves new set of weights W_{m+1} that was used for solving this situation and IM saves the situation X as $S_{m+1} = X$ and $m = m + 1$;

else if $\varepsilon_\alpha \leq \varepsilon_{lim}$, then each μ neuron updates set of weights W_α and the record α in the IM is updated $s_\alpha = X$.

STEP 10. END of the algorithm.

6. Computer model and results of the experimental testing of the proposed immune neural network machine learning of autonomous drones for energy efficiency and collision prevention

A computer model was developed, to prove the efficiency of the proposed autonomous drone control system. The implementation takes place in a two-dimensional (2D) plane, with the assumption that all drones are flying at a uniform altitude.

The model involves an autonomous drone team consisting of eight drones with random sizes, assigned to patrol a specific area. Each drone follows a continuous path between two target points, shuttling back and forth from the first to the second point. Throughout the simulations, the number of drones remains constant.

The drones utilize input data such as their speed, trajectory, and distance to anticipate potential collisions with other drones. Based on this information, each drone calculates the movement direction of other drones relative to its own position and identifies potential collision points.

Each drone specifically calculates the collision probability with other drones that fall within its control area, primarily focusing on those located in front and on the right side. This targeted approach reduces the number of computations needed, optimizing the efficiency of the process.

Screenshots of the computer model working are shown in **Figures 3** and **4**.

In **Figure 3**, the drones operate without any motion control, simply taking the shortest path from one point to another. During this movement, they do not perform risk assessments or implement collision prevention strategies. **Figure 4** employs the INN immune neural network for motion control. The suggested unsupervised drone

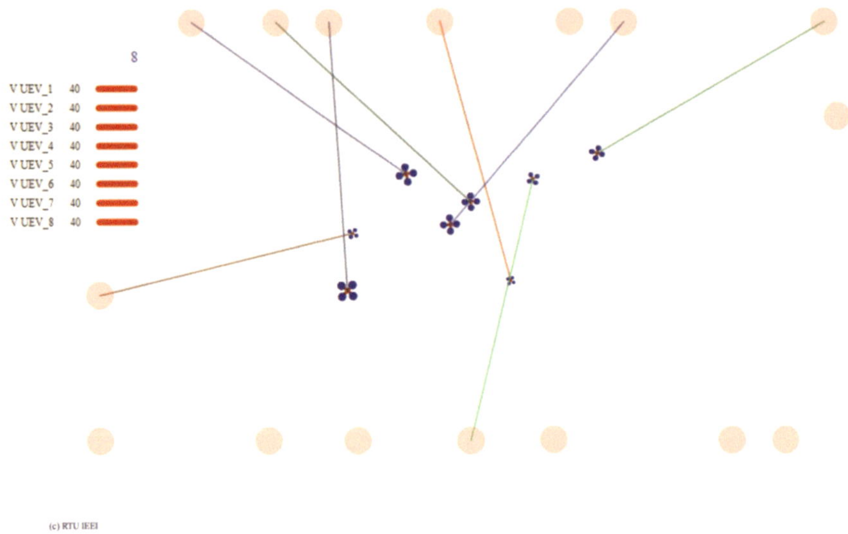

Figure 3.
Computer model during the simulation process without any motion control.

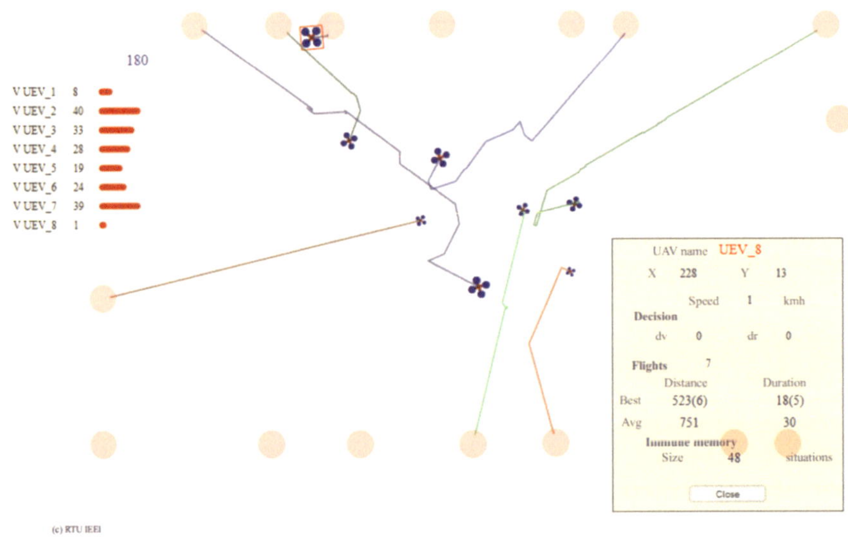

Figure 4.
Computer model during the simulation process with proposed novel INN.

Parameter	Meaning
XYerr	XY position error -XYerr < err < + +XYerr
Verr	speed error -Verr < err < + +Verr
Rerr	rotation error -Rerr < err < + +Rerr
TXdelay	data transmission delay, ms
Sdelay	own positioning data refresh delay, ms
AreaRate	safety zone area rate proportional to the size

Parameter	Meaning
Ap	the magnitude of movement parameter adjustments is determined by the collision probability weight. A higher collision probability weight leads to smaller changes in the movement parameters, while a lower weight results in larger adjustments
Av	speed change weight Av = (1-Ap)/2
Ar	trajectory change weight Ar = (1-Ap)/2
TrLim	maximal number of iterations for the one decision-making process
Sens	collision sensibility
Psens	probability sensitivity
Vlim	maximal speed
Anglim	the maximal possible trajectory change refers to the maximum angle adjustment that can be made to the drone's flight path
Safedist	the maximum distance to the other drone defines the threshold at which the drone begins crash-prevention actions and calculates the collision probability

Table 1.
Parameters of the computer simulation.

Parameter	Meaning
AvDur	Average duration of one flight
SmDur	The shortest duration of the flight
LngDur	The longest duration of the flight
AvDist	Average distance of one flight
SmDist	The smallest distance of the flight
LngDist	The biggest distance of the flight
Trips	Number of complete trips/flights during simulation
Collisions	Number of collisions during simulation

Table 2.
Computer simulation output data.

control system calculates the collision probability and adjusts movement parameters to minimize it. As a result, the paths of the drones in **Figure 4** are not straight, as they are modified to avoid potential collisions.

The database was developed to store parameters of simulation (**Table 1**) and to save the results of the computer simulations (**Table 2**).

The summarized results of the simulations are presented in **Figure 5**. In the next subsections, the description of results is done with reference to the simulation numbers. Better results with the higher number of trips done by drones.

7. Results of simulations without motion control

The first computer simulation was made without any motion control. All the objects were moving straight from one point to another and back again during the simulation.

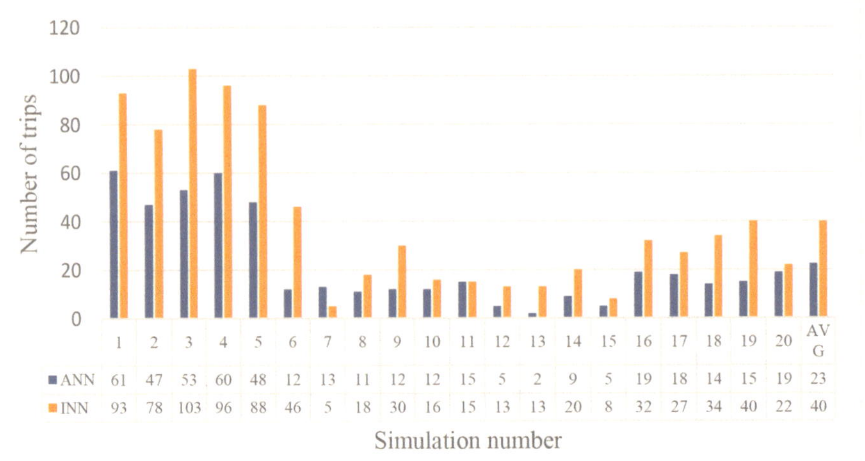

Figure 5.
Comparison of the results of simulations.

ID	XYerr	Verr	Rerr	Txdelay	Sdelay	AreaRate	Ap	Av	Ar	TrLim	Sens	Psens	Vlim	Anglim	Safedist
1	0	0	0	0	0	1	0.8	0.1	0.1	50	6	0.1	70	90	150

Table 3.
Input set 1 of parameters for the computer simulation.

Simulation Nr.	System	AvDur	SmDur	LngDur	AvDist	SmDist	LngDist	Trips	Collisions
AVG 1–5	ANN	22	11	66	640	426	1331	54	0
AVG 1–5	INN	19	11	50	649	425	1397	92	0

Table 4.
Simulation results with the parameter set ID 1.

The results of computer simulation without any motion control show that 8 drones did an average 419 trips and also 60 collisions are detected during 10 minutes (**Figure 5** simulations 1–5).

8. Results of simulations with traditional ANN and new proposed INN without data transmission delays and errors

No data transmission delays and errors were used in these simulations. Other parameters were the same during all simulations (**Table 3**). A description of the input and output parameters is provided in the **Tables 1** and 2.

The output data shows that the number of collisions was reduced to zero during simulations with the ANN, compared to the simulations without any motion control (**Figure 5** simulations 6–10). However, the number of completed trips was much lower (**Table 4**). For the simulations with the proposed INN, the same parameters were used. As a result, no collisions were detected, and the number of trips increased to an average of 92 trips per 10-minute simulation (**Table 4**).

9. Results of simulations with traditional ANN and new proposed INN with data transmission delays and without errors

During these simulations, a data transmission delay of 2000 ms and an own positioning data refresh delay of 500 ms were used (**Figure 5** simulations 6–10).

On average, 12 trips were completed during simulations with a traditional ANN (**Table 5**), which is half as many as during simulations with the proposed INN (**Table 6**). However, collisions were detected in all simulations, indicating that requirements need to be changed and other parameter values must be adjusted to ensure safe driving.

ID	XYerr	Verr	Rerr	Txdelay	Sdelay	AreaRate	Ap	Av	Ar	TrLim	Sens	Psens	Vlim	Anglim	Safedist
2	0	0	0	2000	500	1	0.8	0.1	0.1	50	6	0.1	70	90	150

Table 5.
Input set 2 of parameters for the computer simulation.

Simulation Nr.	System	AvDur	SmDur	LngDur	AvDist	SmDist	LngDist	Trips	Collisions
AVG 6–10	ANN	18	11	61	628	437	1204	12	10
AVG 6–10	INN	42	12	134	761	450	1450	23	8

Table 6.
Simulation results with the parameter set ID 2.

10. Results of simulations with traditional ANN and new proposed INN with data transmission delays, without errors, and with improved parameters

An increased value of the parameter "maximal distance till other drone to start crash prevention and to calculate collision probability" was used during these simulations (**Figure 5** simulations 11–15).

The output data shows that although the number of trips was reduced, they were safer and without collisions. Moreover, the number of trips was twice as large during simulations with the proposed INN (**Tables 7** and **8**).

11. Results of simulations with traditional ANN and new proposed INN with data transmission delays and errors

During these simulations, data delays and errors were introduced to replicate the real-time experiment's conditions (**Figure 5** simulations 15–20).

ID	XYerr	Verr	Rerr	Txdelay	Sdelay	AreaRate	Ap	Av	Ar	TrLim	Sens	Psens	Vlim	Anglim	Safedist
3	0	0	0	2000	500	1	0.8	0.1	0.1	50	6	0.1	70	90	500

Table 7.
Input set 3 of parameters for the computer simulation.

Simulation Nr.	System	AvDur	SmDur	LngDur	AvDist	SmDist	LngDist	Trips	Collisions
AVG 11–15	ANN	30	12	59	773	455	1311	7	0
AVG 11–15	INN	32	11	109	726	439	1569	14	0

Table 8.
Simulation results with the parameter set ID 3.

No collisions were detected in any of the simulations. Furthermore, the number of trips completed was almost twice as high during simulations with the proposed INN (**Tables 9** and **10**).

The example of results for energy efficiency control INN is presented in **Figure 6**.

Initially, the consumption per meter is higher, but during the self-learning of the immune neural network for energy control the energy consumption per meter is significantly reduced, by choosing the best values of control signals when no risks of collisions are detected.

ID	XYerr	Verr	Rerr	Txdelay	Sdelay	AreaRate	Ap	Av	Ar	TrLim	Sens	Psens	Vlim	Anglim	Safedist
4	10	1	1	100	100	1	0.8	0.1	0.1	100	6	0.5	70	90	500

Table 9.
Input set 4 of parameters for the computer simulation.

Simulation Nr.	System	AvDur	SmDur	LngDur	AvDist	SmDist	LngDist	Trips	Collisions
AVG 16–20	ANN	84	14	398	2118	868	5609	17	0
AVG 16–20	INN	52	11	391	1186	432	6709	31	0

Table 10.
Simulation results with the parameter set ID 4.

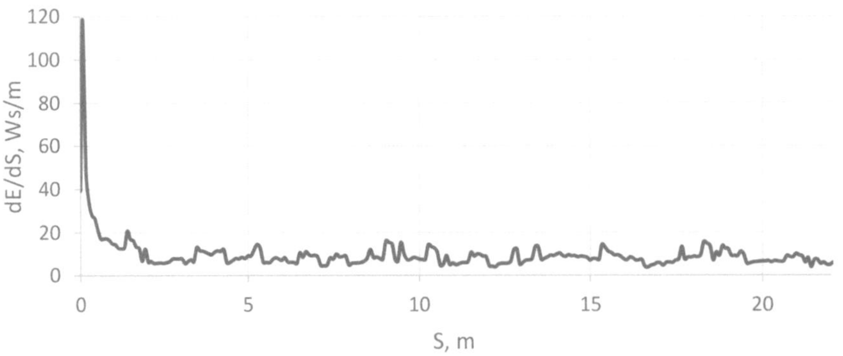

Figure 6.
Example of energy consumption dE per meter dS during the motion.

12. Conclusions

- Centralized, decentralized, and distributed control system models were compared. The results of the comparison show that a distributed system is preferable than a centralized or decentralized one. Distributed models are easier to implement, they have less components, they are cheaper for infrastructure owner, and they are not connected to the specific area. They have also decreased time for reaction and decreased risk at system failure. That is why a distributed system structure was used in this research.

- The researchers developed and provided a description of the system structure proposed in the study. The microcontroller or embedded computer in each drone performs all the necessary functions. It handles tasks like object recognition, risk assessment, opportunity evaluation, and decision-making regarding movement parameter adjustments. Such a solution helps to minimize data processing time because there is no need to transmit the data to the common center and backward.

- In this research, the traditional neural network is included to compare its results with those of the proposed novel INN immune neural network. The objective is to draw conclusions on whether the novel network is better or worse than the traditional one.

- A comparison of ANN- and INN- based algorithms was done, considering the impact on traffic safety and necessary time for decision calculation, where INN presents better results described below. Results approve the proposed hypothesis - an immune neural network can make control decisions to prevent drone collisions with better performance than a traditional neural network in this task.

- Novel immune neural network machine learning is a suitable method for energy efficiency and collision prevention of autonomous drones. Proposed INN does not need to be trained in advance. Collision probability minimization process can be started even with empty immune memory.

- Proposed INN can be used for minimizing the collision probability, improving unsupervised drone safety and faster data processing in real-time conditions with minimal deviation from task performance.

- Proposed INN is better than traditional ANN in drone dangerous situation recognition and prevention task, because of reduced calculation time, which leads to the bigger number of safe trips. Results of computer simulations where drones were able to change their speed and trajectory of motion: without data transmission delays and errors, show that the use of INN helps to increase number of trips by 70% compared to the use of traditional ANN; with data transmission delays and inappropriate maximal distance till other drone to start crash prevention show that the use of INN helps to increase number of trips by 92% compared to the use of traditional ANN and to decrease number of collisions by 25% compared to the use of traditional ANN; with data transmission delays and appropriate maximal distance till other drone to start crash prevention show that the use of INN helps to increase number of trips by 100% compared to the use of traditional ANN; with data transmission delays and errors show that the use of INN helps to increase number of trips by 82% compared to the use of traditional ANN.

- The theme of cybersecurity and loss of signal or communication was not considered in this research. It is considered as a prospect for future scientific research.

- Results of simulations show that the INN reduces the number of iterations and calculation time. It is necessary to analyze whether it will be sufficient for using low-powered systems.

Acknowledgements

This work has been supported by the European Social Fund within Project No 8.2.2.0/20/I/008 "Strengthening of PhD students and academic personnel of Riga Technical University and BA School of Business and Finance in the strategic fields of specialization" of the Specific Objective 8.2.2 "To Strengthen Academic Staff of Higher Education Institutions in Strategic Specialization Areas" of the Operational Program "Growth and Employment".

Author details

Mikhail Gorobetz*, Leonids Ribickis, Anna Beinarovica and Aleksandrs Kornejevs
Riga Technical University, Riga, Latvia

*Address all correspondence to: mihails.gorobecs@rtu.lv

References

[1] SharkSpotter combines AI and drone technology to spot sharks and aid swimmers on Australian beaches [Internet]. 2018. Available from: http://theconversation.com/sharkspotter-combines-ai-and-drone-technology-to-spot-sharks-and-aid-swimmers-on-australian-beaches-92667

[2] Amazon delivered its first customer package by drone [Internet]. 2016. Available from: https://eu.usatoday.com/story/tech/news/2016/12/14/amazon-delivered-its-first-customer-package-drone/95401366/

[3] Safran Group [Internet]. 2023. Available from: https://www.safran-group.com

[4] Ke R, Li Z, Tang J, Pan Z, Wang Y. Real-time traffic flow parameter estimation from UAV video based on ensemble classifier and optical flow. IEEE Transactions on Intelligent Transportation Systems. 2019;**20**(1): 54-64. DOI: 10.1109/TITS.2018.2797697

[5] Rosales C, Gimenez J, Rossomando F, Soria C, Sarcinelli-Filho M, Carelli R. UAVs formation control with dynamic compensation using neuro adaptive SMC. In: Proceedings of the International Conference on Unmanned Aircraft Systems (ICUAS). Atlanta, GA, USA: IEEE; 2019. pp. 93-99. DOI: 10.1109/ICUAS.2019.8798282

[6] Ling L, Niu Y, Zhu H. Lyapunov method-based collision avoidance for UAVs. In: Proceedings of the 27th Chinese Control and Decision Conference (2015 CCDC). Qingdao, China: IEEE; 2015. pp. 4716-4720. DOI: 10.1109/CCDC.2015.7162758

[7] Liu Z, Chen J, Lan F, Xia H. Methodology of hierarchical collision avoidance for high-speed self-driving vehicle based on motion-decoupled extraction of scenarios. IET Intelligent Transport Systems. 2020;**14**(3):172-181. DOI: 10.1049/iet-its.2019.0334

[8] Wang D, Huo C, Tong Z, Yang Y, Wang Y. Research on vehicle anti-collision algorithm based on fuzzy control. In: Proceedings of the 2019 Chinese Control and Decision Conference (CCDC). Nanchang, China: IEEE; 2019. pp. 2361-2366. DOI: 10.1109/CCDC.2019.8833461

[9] Bondarenko A. Development of Knowledge Extraction Methodology from Trained Artificial Neural Networks [Thesis]. Riga: Riga Technical University; 2020

[10] Dung N. Developing models for managing drones in the transportation system in smart cities. Electrical, Control and Communication Engineering. 2019;**15**(2):71-78. DOI: 10.2478/ecce-2019-0010

[11] Gorobecs M, Levchenkovs A. Intelligent electric vehicle motion and crossroad control using genetic algorithms. In: Proceedings of the 10th International Conference on Intelligent Technologies in Logistics and Mechatronics Systems (ITELMS). Lithuania, Panevezys: IEEE; 2015. pp. 1-6

[12] Sell R, Nikitenko A, Ziravecka A, Berkolds K, Vitols K, Czekalski P. Unmanned Electrical Vehicles and Autonomous System Simulation. Riga: RTU Press; 2021. ISBN 978-9934-22-667-0. e-ISBN 978-9934-22-668-7

[13] Haykin S. Neural Networks. A Comprehensive Foundation. Second ed.

Upper Saddle River: Prenctice Hall; 1999 ISBN 8120323734, 9788120323735

[14] Kim DH, Lee KY. Neural networks control by immune network algorithm based auto-weight function tuning. In: Proceedings of the 2002 International Joint Conference on Neural Networks. IJCNN'02 (Cat. No.02CH37290). Honolulu, HI, USA: IEEE; 2002. pp. 1469-1474. DOI: 10.1109/IJCNN.2002.1007734

[15] Mofaddel M, Tavangarian D. A Distributed System with a Centralized Organization. Rostock, Germany; 1997

[16] Graybeal P, Franklin M, Cooper D. Principles of Accounting. Volume 2: Managerial Accounting. OpenStax; 2019

[17] Dewan R, Pahuja N, Kukreja S. Distributed operating system – An overview. International Journal of Research (IJR). 2014;**1**(10):539-548

Chapter 12

Comparison and Transferability of Nitrogen Content Prediction Model Based in Winter Wheat from UAV Multispectral Image Data

Yan Guo, Jia He, Jingyi Huang, Xiuzhong Yang, Zhou Shi, Laigang Wang and Guoqing Zheng

Abstract

Information about the nitrogen dynamic in wheat is important for improving in-season crop precision nutrient management and cultivated land sustainability. To develop unmanned aerial vehicle (UAV)-based spectral models for an accurate and effective assessment of the plant nitrogen content in the key stages (jointing, booting, and filling) of wheat growth, winter wheat experiment plots in Henan Province, China, were used in this study. Based on the K6 multichannel imager, 5-band (Red, Green, Blue, Red edge, and Near-infrared (Nir)) multispectral images were obtained from a UAV system and used to calculate 20 vegetation indices and 40 texture features from different band combinations. Combining the sensitive spectral features and texture features of the nitrogen content of winter wheat plants, BP neural network (BP), random forest (RF), Adaboost, and support vector machine (SVR) machine learning methods were used to construct plant nitrogen content models, and compared for the model performance and transferability. The results showed that the characteristics of different spectral features were different, but most of them had a partial normal distribution. Compared with spectral features, the distribution of texture features was more discrete. Based on Pearson's correlation analysis, 51 spectral and texture features were selected to build four machine learning models. The estimates of plant nitrogen by the RF and Adaboost methods were relatively concentrated, mostly close to the 1:1 line; while the estimates of plant nitrogen from the BP and SVR methods were relatively scattered. The RF method was the best, with coefficient of determination (R^2), root mean square error (RMSE), and mean absolute error (MAE) of 0.811, 4.163, and 2.947 g/m^2, respectively; the SVR method was the worst, with R^2, RMSE, and MAE of 0.663, 5.348, and 3.956 g/m^2, respectively. All models showed strong transferability, especially the RF and Adaboost methods, in predicting winter wheat nitrogen content under rainfed and irrigation water management.

Keywords: spectral feature, texture feature, machine learning, nitrogen, winter wheat, model transferability, UAV

1. Introduction

Wheat is the largest sown area and the most widely distributed food crop in the world [1]. As a major wheat producer, China contributed to more than 17% of the world's total wheat production in 2021 [1, 2]. Stable wheat production is essential to China and the global food supply. Nitrogen is a key and essential nutrient for wheat yield and quality, and accurate information about the nitrogen content of the wheat plant is important for in-season crop growth monitoring, precision fertilizer application, and environmental quality [3–5].

Given the labor-intensive nature of field sampling and laboratory analysis, research has been conducted over the past decades to develop alternative methods for rapid monitoring, and accurate prediction of crop nitrogen content based on spectral imaging collected from near-ground, unmanned aerial vehicle (UAV), and satellite remote sensing systems [6–8]. Particularly, progress has been made in sensitive band screening, vegetation index construction, optimization of prediction and inversion methods, and accuracy improvement [7–10]. However, the efficiency of data collection and performance of various nitrogen prediction models vary with the sensing platforms.

In terms of near-ground sensors, Yang et al. [11] constructed a wheat leaf nitrogen content prediction model based on wheat canopy hyperspectral data collected from a handheld analytical spectral devices (ASD) spectrometer at different growth stages; using the 39 sensitive characteristic bands, the R^2 of the prediction model was as high as 0.998. Similarly, Zhang et al. [12] used ASD wheat canopy hyperspectral data and constructed 14 different vegetation indices such as the Soil Adjusted Vegetation Index (SAVI) to retrieve the nitrogen content of wheat leaves based on the characteristic bands; they found that the combined use of multiple indices could significantly improve the model accuracy compared with a single vegetation index ($R^2 = 0.92$ and 0.83, respectively). With the progress of remote sensing technology, airborne imagers have also been widely used. For example, Li et al. [13] combined handheld canopy spectral data and airborne canopy hyperspectral images at different growth stages of wheat using the N-PROSAIL model to estimate the nitrogen content of wheat canopy and obtained a model R^2 of 0.83; although there was a spectral difference between the handheld ASD and airborne spectrometers, the data fusion did not affect the N-PROSAIL model's performance. The promising prediction results achieved with near-ground and airborne hyperspectral data attract researchers to use the lower-cost UAV remote sensing data for rapid monitoring and inversion of crop nitrogen, but the low spectral resolution of UAV multispectral data affects the prediction performance of nitrogen models. In addition to spectral models based on spectral reflectance, the rich texture information of the UAV multispectral images has not been widely used in constructing plant nitrogen models. Previous studies have shown that the texture feature can improve the identification of useful spatial features from the original images and enhance the inversion accuracy when retrieving crop parameters [14–16]. For example, Jia and Chen [14] established a model using principal component regression analysis for predicting the nitrogen content of winter wheat using UAV image features at a spectral resolution of 0.06 m; the accuracy of the model established by fusing the spectral and texture features ($R^2 = 0.68$) of UAV multispectral images was improved by more than 10% compared with that established by a single vegetation index ($R^2 = 0.66$) or texture feature ($R^2 = 0.65$). Zheng et al. [17] evaluated the potential of integrating texture and spectral information from UAV-based multispectral imagery for improving the quantification of nitrogen status in rice crops, indicating model vegetation indices (Vis) with a R^2 of 0.14, texture features with a R^2 of 0.41, and the combination of VIs and texture features with a R^2 of 0.68. The results revealed the potential of image textures derived from UAV images for estimation of winter wheat nitrogen status. Therefore, a comprehensive analysis of the sensitivity of spectral and texture features to crop nitrogen content, and the predictive model establishment using appropriate methods are of great significance to improve the accuracy of nitrogen content prediction, enhance the applicability of models, and reduce costs.

The methods for prediction and inversion of plant nitrogen content mainly include statistical and physical methods. Statistical methods used univariate and multiple regression to establish linear, logarithmic, and power function models [14, 18]. Physical models are mainly radiative transfer models and geometric optical models, and the crop nitrogen content predicted by screening feature bands through sensitive parameter analysis, using lookup table method and artificial neural network (ANN) method [8, 13, 19]. In recent years, with the advancement of data mining techniques, various programs, such as artificial neural networks (ANN), genetic algorithms (GA), random forest (RF), and other hybrid methods, have been increasingly applied to the prediction of crop nitrogen content, which outperform traditional models in terms of accuracy [9, 20, 21]. For example, Chlingaryan et al. [20] compared the analysis of traditional statistical analysis methods and machine learning methods in crop nitrogen content and yield prediction. They found that machine learning regression methods, such as least squares support vector machines (LS-SVR) and back-propagation neural networks (BPNN), have promised a higher accuracy. In addition, different machine learning methods differ in prediction accuracy [22, 23]. For example, Yang et al. [11] used backward transmission neural network (BP), SVR, and radial basis neural network (RBF) methods for the prediction of canopy nitrogen in winter wheat with model R^2 of 0.82 (SVR) to 0.98 (RBF); Qiu et al. [22] used Adaboost, ANN, K-neighborhood (KNN), partial least squares (PLSR), RF, and SVR machine learning regression methods for the prediction of rice nitrogen nutrient index, and found that the RF achieved the highest model accuracy with a R^2 of 0.98 during the filling stage. Although machine learning methods are proven to be superior to traditional statistical analysis methods, few studies investigated the transferability of the models when applied to UAV images. To explore the model prediction effects and transferability of general machine learning methods (BP and SVR) and ensemble learning methods (RF and Adaboost) for constructing plant nitrogen content models, especially in agricultural applications, this study aims to compare the performance of different machine learning models in predicting the wheat nitrogen content with a combination of spectral and texture features and provide insights for future model deployment to support the rapid in-season assessment of nitrogen nutrition and precision fertilization across large extents. Specifically, our study will focus on the following aspects:

1. Extracting spectral and texture features of UAV images based on the acquired multispectral images at key growth stages of winter wheat.

2. Feature optimization by Pearson's correlation analysis using the extracted spectral and texture features.

3. Comparing the model accuracy for estimating wheat canopy nitrogen content using different machine learning methods (BP, RF, Adaboost, and SVR).

4. Evaluation of the computation efficiency and transferability of different machine learning methods under rainfed and irrigation water management.

2. Materials and methods

2.1 Study area and experimental design

The study area is located in Shangshui County, Henan, China, as shown in **Figure 1**. The study area is flat and belongs to the temperate continental monsoon climate with cold and dry winters and high temperatures and rainy summers. Winter wheat, maize, cotton, and other crops are mainly cultivated. The growth cycle of winter wheat is 8 months. Generally, it is sown in October and harvested at the end of May and early June of the following year. The experiment was conducted in the jointing, booting, and filling stages of winter wheat, and the field cover did not change much during these three phenological periods. Field activities affecting the ground surface, such as plowing and sowing, were avoided in the experimental plots. The soil type in the study area is sandy ginger black soil (Chinese Soil Taxonomy Classification).

Five nitrogen application levels were set using a randomized block experiment design, which were N0 (0), N6 (60 kg/hm^2), N12 (120 kg/hm^2), N18 (180 kg/hm^2), and N24 (240 kg/hm^2), and of which 50% was applied as base fertilizer and the remaining 50% was applied at the jointing stage. All treatments had 150 and 90 kg/hm^2

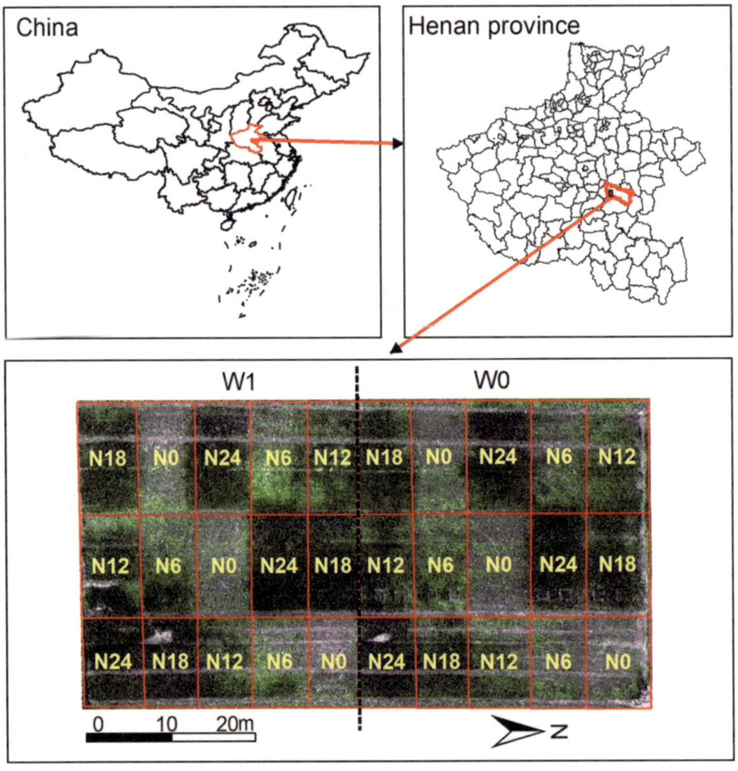

Figure 1.
Location of the study area and the spatial distribution of the experimental design.

of phosphorus and potassium fertilizer, all of which were applied as base fertilizer. Two levels of water set were: rainfed only (W0) and normal irrigation (W1). Each nitrogen level treatment was repeated three times, so a total of 30 plots. The spatial distribution of the experimental design is shown in **Figure 1**.

2.2 Data acquisition and processing

2.2.1 UAV-based multispectral image acquisition and preprocessing

This study employed a UAV system consisting of a six-rotor DJI M600 UAV and a K6 multichannel multispectral imager (Anzhou Technology, Beijing, Co., Ltd.) to acquire multispectral images. The K6 multispectral imager mounted onboard the UAV had an incident light sensor and five spectral bands with center wavelengths at Blue (450 nm), Green (550 nm), Red (685 nm), Red edge (725 nm), and Near-infrared (780 nm, Nir). During the growth period of winter wheat from 2020 to 2022, the multispectral images of the winter wheat canopy were acquired at the jointing, booting, and filling stages with a flight height of 50 m. When the aircraft flew, the lens was vertically downward with a field angle of 30° and a heading overlap of 70%, and a lateral overlap of 75%. The canopy reflectance data of winter wheat were extracted by format conversion, image mosaicing, geographic information correction, and radiometric calibration. The digital number (DN) values of the images were transformed into reflectance values per band by applying the empirical line model derived from the measured reflectance values and DN values of the five calibration canvases.

2.2.2 Ground data acquisition and processing

The acquisition of ground-truth data is synchronized with the acquisition of the UAV multispectral images. The area with uniform growth is selected in every plot, and 20 of the single stem samples were taken in sealed bags by fixing the total number of stems in a 1-meter double row. The plant organs were separated into leaves, stems, and ears in the laboratory and placed in paper bags respectively. They were killed at 105°C and dried to a constant weight at 80°C. After the organs were crushed, the nitrogen content was determined by the Kjeldahl method. Finally, 360 measured nitrogen values of winter wheat plant nitrogen content were obtained for the three growth stages. The samples were divided into training and test datasets according to the 1:1 ratio, and the models were trained using cross-validation.

2.3 Methods

2.3.1 Feature extraction

1. Spectral vegetation index calculation

Since the launch of the Earth Resources Satellite, scientists have begun to study the relationship between spectral response and vegetation [24–27]. Crops with nitrogen deficiency will show obvious apparent characteristics such as reduced coverage and yellowing of leaves [28, 29]. This study selected 20 common vegetation indices used for nitrogen content prediction, and the formulas are shown in **Table 1**.

2. Texture feature extraction

Vegetation indices	Abbreviation	Calculation formulas	References
Green Normalized Vegetation Index	GNDVI	$(R_{nir}\text{-}R_{green})/(R_{nir} + R_{green})$	[30]
Green Optimized Soil Adjusted Vegetation Index	GOSAVI	$1.16 \times ((R_{nir}\text{-}R_{green})/(R_{nir} + R_{green} + 0.16))$	[31]
Normalized Difference Vegetation Index	NDVI	$(R_{nir}\text{-}R_{red})/(R_{nir} + R_{red})$	[32]
Modified Simple Ratio Index	MSR	$((R_{nir}/R_{red})\text{-}1)/(((R_{nir}/R_{red}) + 1)^{0.5})$	[33]
Rededge Optimized Soil Adjusted Vegetation Index	REOSAVI	$1.16 \times ((R_{nir}\text{-}R_{red})/(R_{nir} + R_{red} + 0.16))$	[34]
Rededge Renormalized Difference Vegetation Index	RERDVI	$(R_{nir}\text{-}R_{red\ edge})/(R_{nir} + R_{rededge})^{0.5}$	[30]
Chlorophyll Absorption Ratio Index	CARI	$(R_{red\ edge}\text{-}R_{red})\text{-}0.2 \times (R_{rededge} + R_{red})$	[35]
Optimized Soil Adjusted Vegetation Index	OSAVI	$(R_{nir}\text{-}R_{red})/(R_{nir} + R_{red} + 0.16)$	[35]
Normalized Green-Blue Difference Index	NGBDI	$(R_{green}\text{-}R_{blue})/(R_{green} + R_{blue})$	[36]
Enhanced Vegetation Index	EVI	$2.5 \times ((R_{nir}\text{-}R_{red})/(R_{nir} + 6 \times R_{red}\text{-}7.5 \times R_{blue}+1))$	[37]
Triangular Vegetation Index	TVI	$0.5 \times (120 \times (R_{nir}\text{-}R_{green})\text{-}200 \times (R_{red}\text{-}R_{green}))$	[38]
Atmospherically Resistant Vegetation Index	ARVI	$(R_{green}\text{-}R_{red})/(R_{green} + R_{red}\text{-}R_{blue})$	[39]
Excess Green Index	EXG	$2 \times R_{green}\text{-}R_{red}\text{-}R_{blue}$	[40]
Ratio Vegetation Index	RVI	R_{nir}/R_{red}	[41]
Modified Triangular Vegetation Index	MTVI	$(1.5 \times (1.2 \times (R_{nir}\text{-}R_{green})\ 2.5 \times (R_{red}\text{-}R_{green})))/(((2 \times R_{nir} + 1)^2\text{–}6 \times R_{nir}\text{-}5 \times (R_{red})^{0.5}\text{–}0.5)^{0.5})$	[42]
Soil Adjusted Vegetation Index	SAVI	$1.5 \times (R_{nir}\text{-}R_{red})/(R_{nir} + R_{red} + 0.5)$	[43]
Normalized Blue-Green Difference Vegetation Index	GBNDVI	$R_{nir}\text{-}(R_{green} + R_{blue})/(R_{nir} + (R_{green} + R_{blue}))$	[30]
Renormalized Difference Vegetation Index	RDVI	$(R_{nir}\text{-}R_{red})(R_{nir} + R_{red})^{0.5}$	[42]
Difference Vegetation Index	DVI	$R_{nir}\text{-}R_{red}$	[44]
Optimized Vegetation Index	VIplot	$1.45 \times (R^2_{nir} + 1)(R_{red} + 0.45)$	[45]

Table 1.
Vegetation indices and the calculation formulas.

Texture feature extraction methods mainly include statistical methods, such as gray level cooccurrence matrix (GLCM), texture spectrum and geometric methods; model methods of random field model and fractal model methods; signal processing methods and structural analysis methods [46, 47]. Among these methods, the GLCM method is an image recognition technology currently recognized by the academic community as an image recognition technique with strong robustness and adaptation characteristics, which can effectively achieve the classification and retrieval of images and maximize the accuracy of remote sensing image classification processing [16, 26]. In this study, the texture features from five bands in multispectral images are extracted through the GLCM method, and the extracted texture feature information mainly includes eight indicators of con, cor, dis, ent, hom, mean, sm, and var. The specific calculation method is described in Zhou et al. [26].

2.3.2 Machine learning regression method

1. Back-propagation neural network

Back-propagation neural network (BP) is a multilayer feedforward network using error back-propagation for model training, which is one of the most widely used neural network models [48, 49]. This study uses identity as the plant nitrogen content training activation function of the model. Also, to prevent overfitting, parameters such as learning rate and regularization are often introduced to optimize the model [50]. In this study, a 3-layer network structure was used, and a quasi-Newtonian method family optimizer (lbfgs) was used to improve the running speed. In this study, the detailed parameters for BP are shown in **Table 2**.

2. Random forest

Random forest is a typical representative of ensemble learning with bagging idea, a supervised machine learning method constructed by integration with decision tree as the base learner, and it introduces randomness in the training process of decision trees to make it have excellent resistance to overfitting as well as noise resistance; moreover, RF can be trained in parallel during model training to improve the efficiency of training, while feature importance can be obtained [51, 52]. RF reflects its randomness from two aspects: sample selection and feature selection. Combined with the study of Liepe et al. [53], the parameters of RF node splitting evaluation criterion and maximum depth of the tree in this study are set in **Table 2** after several runs and the algorithm steps are as follows:

① Draw the training dataset from the original sample dataset. In each round, n training samples (with put-back sampling) are drawn from the original sample dataset using the Bootstrap method. A total of k rounds are performed and k training sets are obtained.

② Each time a training dataset is used to get a model, k training datasets get a total of k models.

③ For the classification problem: the k models obtained in the previous step are voted to obtain the classification results, and the mean value of the above models is calculated as the final result.

3. Adaboost

Adaboost is the abbreviation of "adaptive boosting," which was proposed by Yoav Freund and Robert Schapire in 1995. It is a typical representative of the idea of

BP		RF		Adaboost		SVR	
Parameters	**Parameter value**	**Parameters**	**Parameter value**	**Parameters**	**Parameter value**	**Parameters**	**Parameter value**
Data cut	0.5	Data cut	0.5	Data cut	0.5	Data cut	0.5
Data shuffling	Yes	Data shuffling	Yes	Data shuffling	Yes	Data shuffling	Yes
Cross-validation	30% off validation	Cross-validation	30% off validation	Cross-validation	30% off validation	Cross-validation	30% off validation
Activation function	identity	Identity node split evaluation criterion	mse	Number of base classifiers	100	Penalty factor	1
Solver	lbfgs	Minimum number of samples for internal node splitting	2	Loss function	linear	Kernel function	linear
Learning rate	0.1	Minimum number of samples of leaf nodes	1	Base classifier	Decision tree classifier scale	Kernel function coefficient	scale
L2 regular term	1	Maximum depth of tree	10	Learning rate	1	Maximum number of terms in kernel function	3
Number of iterations	1000	Maximum number of leaf ncdes	50	/	/	Error convergence condition	0.001
Number of hidden layers 1st neurons	100	Number of decision trees	100	/	/	Maximum number of iterations	1000

Table 2.
Parameters for BP, random forest (RF), Adaboost, and support vector machine (SVR) methods.

ensemble learning with boosting idea, and the operation is performed by continuous iterations, adding a new weak learner in each round until some predefined sufficiently small error rate is reached. Adaboost method is very sensitive to noisy and abnormal data and is less prone to overfitting than most other learning algorithms. The Adaboost method operates by iterating continuously, adding a new weak classifier in each round, until a predetermined small enough error rate is reached. Each training sample is given a weight indicating the probability of being selected by a classifier into the training set, and the weights are continuously adjusted so that the Adaboost method can "focus" on those samples that are difficult to distinguish [54–56]. The detailed parameters for Adaboost are shown in **Table 2**.

4. Support vector machine regression

Support vector machine regression (SVR) characterizes data into a high-dimensional data feature space using a nonlinear mapping, so that the independent and dependent variables have good linear regression characteristics in the high-dimensional data feature space, which is fitted in that feature space and then returned to the original space [57, 58]. The details can be shown in the **Figure 2**. Given the training sample $D = \{(x_1,y_1),(x_2,y_2),\ldots,(x_n,y_n)\}$, it is desired to learn an $f(x)$ such that it is as close as possible to y. w and b are the parameters to be determined. In this model, the loss is zero only when $f(x)$ is identical to y. The SVR method assumes that we can tolerate a deviation of at most ε between $f(x)$ and y. The loss is calculated when and only when the absolute value of the difference between $f(x)$ and y is greater than ε. This is equivalent to constructing an interval band of width 2ε centered on $f(x)$, and if the training sample falls into this interval band, it is considered to be predicted correctly.

In regard to the kernel functions, linear, polynomial, radial basis, sigmoid, etc., are commonly used [50, 59]. Among them, the linear kernel function has the advantages of high efficiency and wide range of applications, combined with the study of Yi et al. [60], the linear kernel function was selected in this study to meet the demand and also improve efficiency, and other parameter settings are shown in **Table 2**.

2.3.3 Model evaluation metrics

Root mean square error (RMSE), mean absolute error (MAE), and coefficient of determination (R^2) are used to measure the prediction effect of nitrogen content in

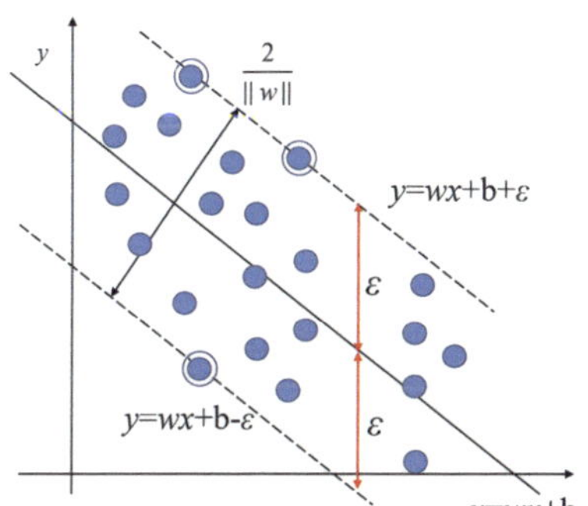

Figure 2.
Schematic diagram of support vector machine (SVM) regression.

winter wheat plants. The smaller the values of RMSE and MAE, the more accurate the model is. R^2 compares the predicted values with the case where only the mean value is used. The closer the result is to 1, the more accurate the model is. The RMSE, MAE, and R^2 are calculated according to Bellis et al. [61].

3. Results

3.1 Statistical analysis of UAV multispectral image data features

3.1.1 Spectral features

The physiological and biochemical parameters of winter wheat plants differed across the spectral features as shown in **Figure 3**. The skewness and kurtosis of the Nir band reflectance were smaller than zero, showing a left-skewed and flat broad peaks distribution; while the skewness coefficient and kurtosis coefficient of the Red edge, Red and Blue bands' reflectance were larger than zero, showing a right-skewed and sharp peaks distribution. The skewness of the Green band reflectance was less than zero and the kurtosis coefficient was greater than zero, showing a left-skewed and flat broad peaks distribution. Combined with the magnitude of the kurtosis coefficient and skewness coefficient values, there were differences in the distribution of reflectance data in different bands, indicating the different nitrogen content or other growth parameters of winter wheat plants.

The 20 vegetation indices constructed based on spectral reflectance, the mean, standard deviation (SD), skewness coefficient, and kurtosis coefficient of TVI, RVI and MSR were greater than the others, and the coefficient of variation (the ratio of SD to mean) values were mostly around 0.20. From the data distribution characteristics, the skewness coefficient and kurtosis coefficient of VIplot, DVI, GOSAVI, and

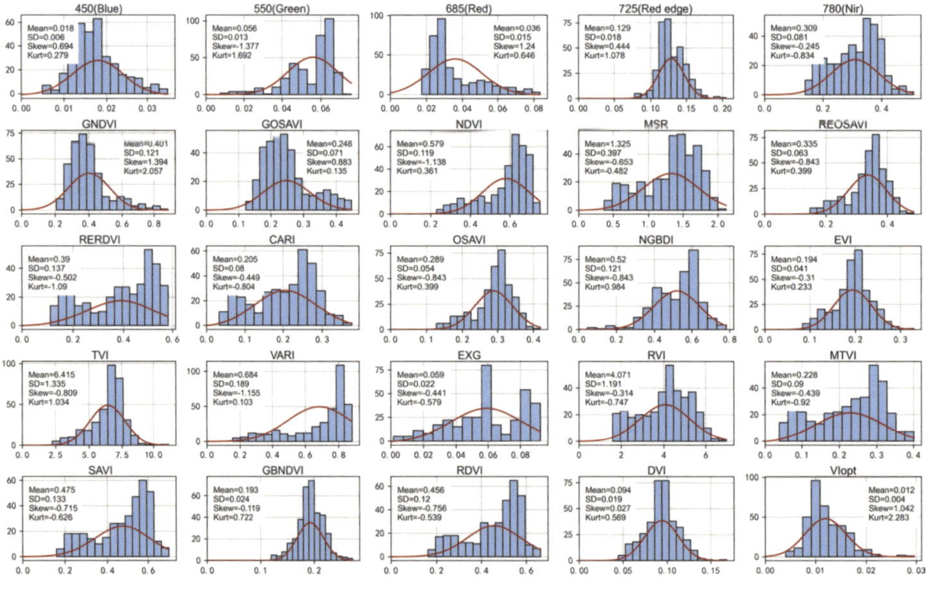

Figure 3.
Statistical analysis of spectral features.

GNDVI were all greater than zero, showing a right-skewed and sharp peaks distribution; the skewness coefficient and kurtosis coefficients of RDVI, SAVI, MTVI, RVI, EXG, CARI, RERDVI, and MSR were less than zero, showing a left-skewed and flat broad peak; the skewness coefficients and kurtosis coefficients of GBNDVI, VARI, TVI, EVI, NGBDI, OSAVI, REOSAVI, and NDVI were less than zero and the kurtosis coefficients were greater than zero, showing the distribution of left-skewed and sharp peaks. In summary, different spectral features had different numerical magnitudes and distributions, but most of them show close to skew-normal distributions, indicating the similarities and dissimilarities in the response of these spectral features to wheat nitrogen content, which can provide a basis for the construction of nitrogen content model using vegetation indices.

3.1.2 Texture features

Texture features are inherent attributes of remote sensing images, especially for the high spatial resolution UAV data. The basic characteristics of these 40 texture features were analyzed and are shown in **Figure 4**.

The texture features of con, dis, ent, mean, and var. increased with increasing wavelength, while hom and sm decreased with increasing wavelength. Compared with the spectral feature values, the difference between the maximum, minimum, and median values of texture features was larger, which had a differential amplification effect on the construction of the nitrogen content model for winter wheat. From the perspective of data distribution characteristics, the skewness coefficient and kurtosis coefficient of 12 texture feature values, including var_780, var_685, var_550, var_450, sm_780, sm_725, hom_780, dis_685, cor_685, con_685, con_550, and con_450, were all greater than zero, showing a right-skewed and sharp peaks distribution; the skewness coefficient and kurtosis coefficient of sm_685, ent_725, ent_550, ent_450, hom_550, dis_450, cor_780, and mean_780 were less than zero, showing a left-skewed and flat broad peaks distribution; hom_725, mean_725, dis_725, sm_450, cor_450, mean_685, mean_450, hom_450, dis_780, sm_550, var_725, con_725, dis_550, ent_685, cor_725, cor_550, and con_780 texture eigenvalues had skewness coefficients larger than zero and kurtosis coefficients smaller than zero, showing a right-skewed and flat broad peaks distribution; 685hom and 780ent had skewness coefficients smaller than zero and kurtosis coefficients larger than zero, showing a left-skewed and sharp peaks distribution. Compared with the spectral features, the texture eigenvalues were more discrete and the coefficient of variation (CV) was also larger. This dispersion provides a basis for the construction of an accurate nitrogen content model for winter wheat.

3.2 Sensitive analysis of the feature responses to the nitrogen content

To screen out the sensitive characteristics of plant nitrogen content, Pearson's correlation analysis was conducted between 25 spectral features and 40 texture features and the measured nitrogen content, and the results are shown in **Figure 5**. For spectral features, 25 features passed the 0.01 extremely significant level test, except for three features of 550-nm band reflectance, GNDVI, and GOSAVI; for texture features, except for 11 features, mean_450, var_725, sm_725, mean_725, hom_725, ent_725, dis_725, con_725, cor_685, and dis_780, whereas the other 29 texture features passed the 0.01 extremely significant level test. Specifically, some correlation coefficients (r) had absolute values greater than 0.5, such as 17 species

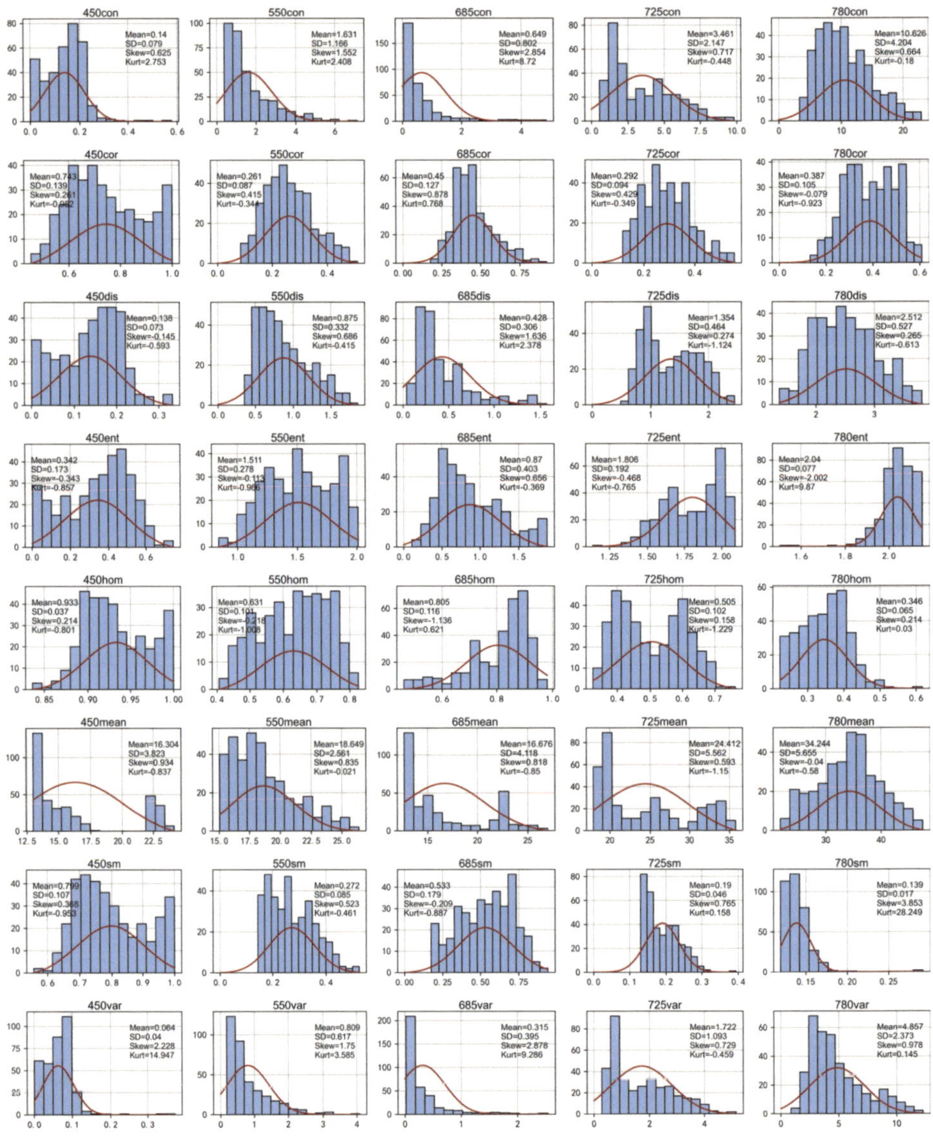

Figure 4.
Statistical analysis of texture features.

including spectral reflectance of 450-nm, 685-nm, 780-nm bands and vegetation indices of TVI, VARI, REOSAVI, OSAVI, NDRGI, RVI, MSR, NDVI, VARI, CARI, RDVI, SAVI, MTVI, RERDVI, and 4 texture features, including cor_550, cor_725, cor_725, and mean_780. Moreover, there was a positive correlation between spectral features and wheat nitrogen content, and a negative correlation between texture features and plant nitrogen content. Therefore, to retain the sensitive characteristics of plant nitrogen content as much as possible, the 51 spectral features and texture features that passed the 0.01 extremely significant level test were taken as input variables for the construction of the plant nitrogen content prediction model in the next step.

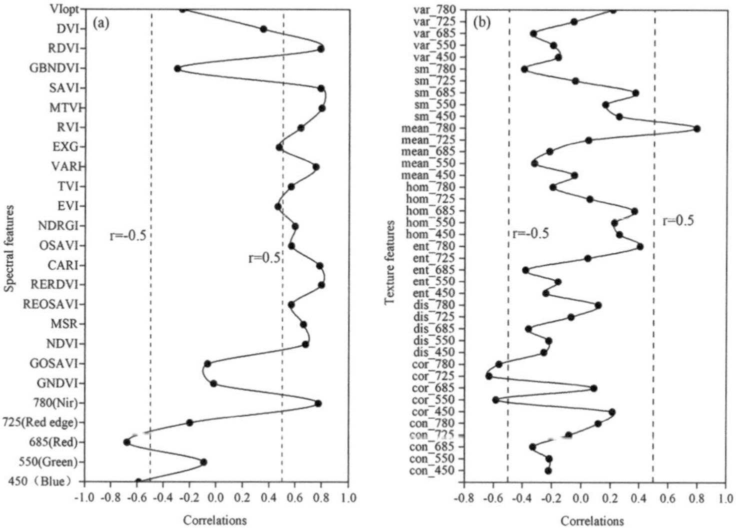

Figure 5.
Correlation coefficients between spectral and texture features and nitrogen content in winter wheat plants.

3.3 Prediction of nitrogen content in winter wheat plant based on machine learning methods

Based on the 51 spectral features and texture features, BP, RF, Adaboost, and SVR regression methods were used to predict the nitrogen content in winter wheat plants. The model was first trained based on the training dataset and then evaluated using the test dataset. The measured and predicted nitrogen for the test dataset are shown in **Figure 6**. The machine learning models had different effects on the prediction of winter wheat plant content. From the 95% confidence interval (CI), the confidence interval span between the measured and predicted nitrogen values of BP and SVR methods was larger than that of RF and Adaboost methods, and the data distribution was more scattered, while the measured and predicted nitrogen values of RF and Adaboost methods were more close to the 1:1 line.

4. Discussion

4.1 Training efficiency of different machine learning methods

Under the same environmental conditions such as data segmentation, shuffling methods, cross validation, etc., as shown in **Table 2**, the training time of the model varies greatly. Using a computer with Intel Core i7-9700K CPU and 64 GB RAM, the SVR method takes the shortest time of 0.022 s, the RF and Adaboost methods take time of 0.778 s and 0.831 s, respectively, and the longest time is the BP method (3.122 s), which is 142 times longer than that of the SVR method. This is consistent with the results obtained in other studies, such as in Jeung et al. [51], Du et al. [48], Fernández Habas et al. [52], and Lin and Liu [62], on the prediction of flow scouring efficiency, thermal efficiency, pasture quality, and soil total nitrogen using these machine learning methods.

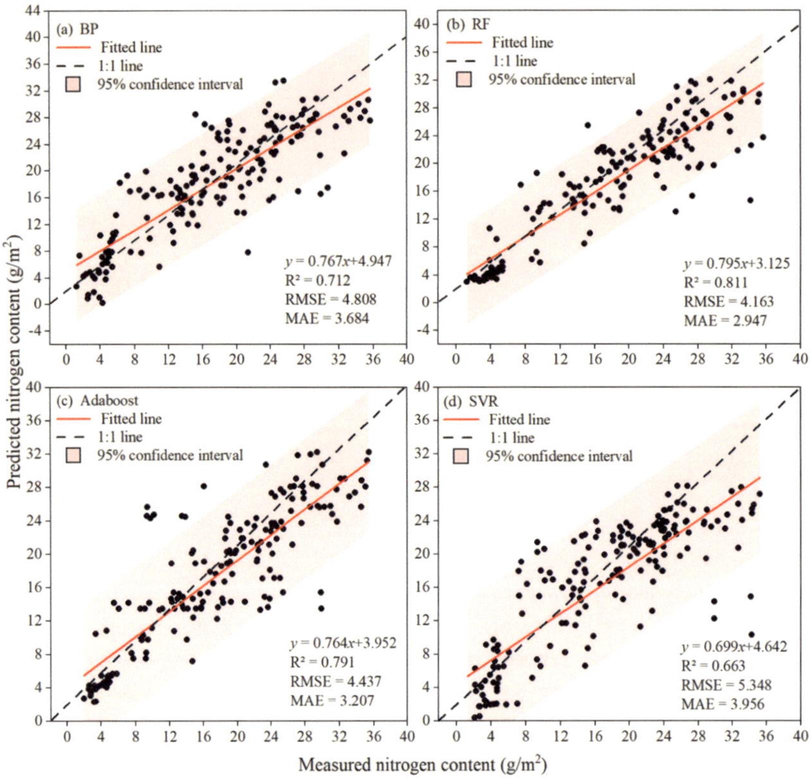

Figure 6.
Relationship between the predicted and measured nitrogen content in a winter wheat plant with the four different machine learning methods.

4.2 Effects of different machine learning methods on the nitrogen content prediction models

The model evaluation metrics for the training and test datasets from the four methods are shown in **Table 3**. In terms of the training dataset, the R^2 of the different models follow the order of Adaboost (0.988), RF (0.964), BP (0.837), and SVR (0.703); the values of RMSE and MAE were similar to that of the R^2. However, this is not the case for the test dataset. For R^2, the best model is RF (0.811), followed by Adaboost (0.791), BP (0.712), and SVR (0.663). This indicates that the Adaboost model has been overfitted compared to RF, BP, and SVR.

Furthermore, the model evaluation was conducted using the test datasets for the constructed model, and the relationship between the measured and predicted nitrogen values of test dataset was shown as per sample given in **Figure 7**. The local fitting effects of the four methods have good performance, which were mainly related to the distribution of plant nitrogen content data characteristics. In this study, the nitrogen content values were mainly concentrated in the range of 12–28 g/m^2, and the trained model had better prediction ability for the local values in this range, so the agreement between the measured and predicted nitrogen is high for the test datasets; overall, there was a trend of underestimation in the high nitrogen content area, while the estimation was relatively good in the low nitrogen content area. Along with the over−/−underprediction of different models demonstrated earlier, the results indicate

Methods	Datasets	RMSE (g/m^2)	MAE (g/m^2)	R^2
BP	Training dataset	3.955	3.040	0.837
	Test dataset	4.808	3.684	0.712
RF	Training dataset	1.749	1.299	0.964
	Test dataset	4.163	2.947	0.811
Adaboost	Training dataset	0.202	0.067	0.988
	Test dataset	4.437	3.207	0.791
SVR	Training dataset	5.210	4.047	0.703
	Test dataset	5.348	3.956	0.663

Table 3.
Evaluation of the models constructed by random forest (RF), Adaboost, and support vector machine (SVR) methods.

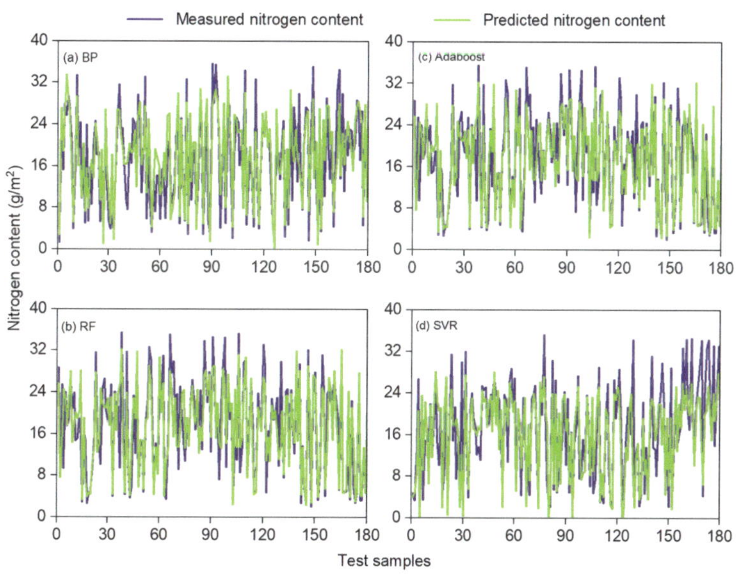

Figure 7.
Curve fitting effects of the test datasets.

that the efficiency and accuracy of the prediction models constructed by the RF and Adaboost methods are outstanding, which is inextricably linked to the principles of the algorithms.

4.3 Transferability of the prediction model of nitrogen content in winter wheat plants

In this study, BP, RF, Adaboost, and SVR machine learning methods were used to construct the nitrogen prediction model for winter wheat, and good prediction results were achieved when both water management (rainfed and irrigation) treatments were combined. How about the transferability of the established model across different water treatments? This is an important issue regarding the generalizability of the model [63], which has practical implications for researchers who want to apply the models developed from one water management regime to another. To answer this question, the models constructed by different machine learning methods using datasets under W0 or W1 water treatments were evaluated. For example, models were first fitted using datasets from W1 treatment only and predicted onto the W0 treatment plots and vice versa (**Figures 8** and **9**). The prediction effects of the four methods on the nitrogen content of W0 and W1 treatments trained using W1 and W0 treatments were the same as those trained using both W0 and W1 datasets, both of which were closer to the 1:1 line for the RF and Adaboost methods. The R^2 of transfer prediction results for the models constructed by BP, RF, Adaboost, and SVR methods were 0.751, 0.723, 0.720, and 0.660 for the prediction of nitrogen content in W0 treatment and 0.512, 0.693, 0.612 (trained using data from W1 treatment) and 0.452

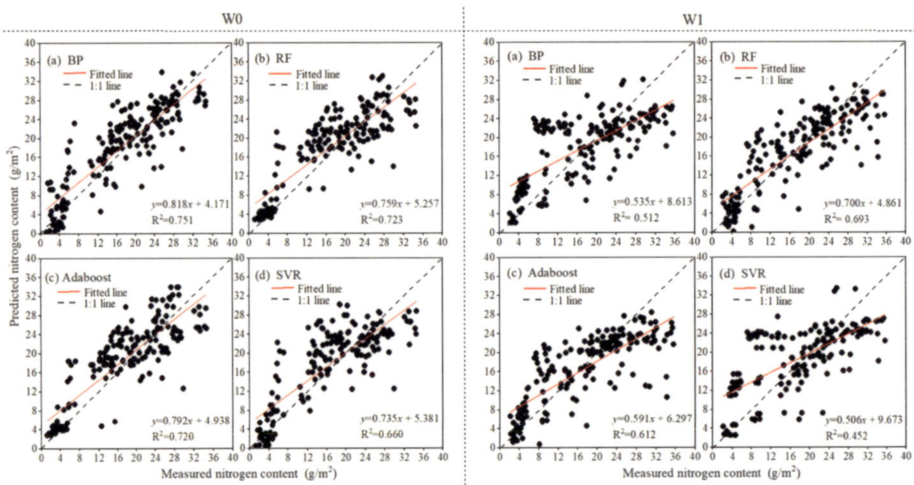

Figure 8.
Transferability of models constructed by the four machine learning methods under rainfed (W0) and irrigation (W1) treatments.

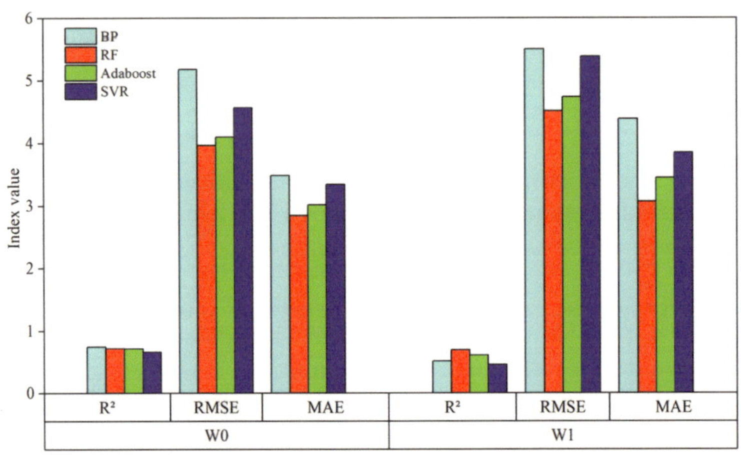

Figure 9.
Comparison of the transferability of models constructed by the four machine learning methods for rainfed (W0) and irrigation (W1) treatments.

for the prediction of nitrogen content in W1 treatment (trained using data from W0 treatment), respectively. This is also the case for the RMSE and MAE. As a result, the transfer prediction ability of the plant nitrogen content prediction model constructed by RF and Adaboost methods was better than that of BP and SVR methods. Although the nitrogen content prediction model constructed in this study has good local transferability, future research is needed to test whether such models could be transferred to other wheat production areas or even to other crops.

4.4 Mechanism analysis of the four machine learning methods for nitrogen content prediction

Machine learning is a data-driven method, which can achieve accurate prediction by fully mining the information in the dataset. Machine learning has become a research hotspot for prediction in many disciplines [48, 51, 52, 62–66]. However, there are differences in the design of the machine learning methods, and this study focuses on the influence of BP, RF, Adaboost, and SVR models on the prediction of nitrogen content in winter wheat. The BP method has a relatively strong learning ability, but it requires more parameters to be fitted; the model training also takes a long time, and different solvers and activation functions may affect the efficiency of the model. The training of RF and Adaboost methods is adjustable with relatively simple parameters and fast fitting speed. The SVR method can solve high-dimensional problems with strong generalization ability and relatively low dependence on the overall data, but it is difficult to determine the appropriate kernel function. Therefore, there is a tradeoff between the efficiency and accuracy of the machine learning models.

In the study, the RF and Adaboost methods are more prominent with the R^2 above 0.8, mainly because both methods belong to ensemble learning based on the idea of bagging and boosting, respectively. During processing of the models' construction, a number of learners are combined to get a new learner, so as to achieve a better learning effect, which fully reflects the "group wisdom" of machine learning. In addition, both methods are extracted from the original dataset using the Bootstrap strategy and reorganized to form a subset as large as the original dataset. This means that samples inside the same subset can be recurring, and samples in different subsets can also be recurring. Moreover, unlike a single decision tree that selects an optimal feature to segment nodes after considering all features in the segmentation process, the RF method selects the optimal feature variable among these features by randomly examining certain feature variables in the base learner, similar to "democratic voting," and this randomness makes the generalization ability and learning ability of the RF model superior to those of the individual learner. This performance has also been verified in the literature [49, 52, 53, 64]. The Adaboost method takes into account the weights of each classifier in the sampling process, similar to "elite selection," but if the data are not balanced, the model accuracy decreases [57]. Comprehensively, the RF and Adaboost methods are more effective for plant nitrogen content prediction models when considered together.

5. Conclusions

Based on the 5-band multispectral reflectance acquired by the K6 multichannel multispectral imager, 20 vegetation indices and 40 texture features were obtained by computational analysis. Different spectral features had different numerical magnitudes and distribution characteristics with an approximate skew-normal distribution, while the texture features were more discrete compared with the spectral features. In total, 51 spectral features and texture features that passed the 0.01 significant level test were selected to construct models using the BP, RF, Adaboost, and SVR methods with validation R^2 of 0.712, 0.811, 0.791, and 0.663, respectively. The RF and SVR methods tend to underestimate the wheat nitrogen content, while BP and Adaboost slightly overestimated the wheat nitrogen content. When predicting the nitrogen content in winter wheat under different water treatments, the model shows a strong transferability, especially the RF and Adaboost methods. Combining R^2, RMSE, and MAE, the RF and Adaboost methods have better computation time, accuracy, and transferability for nitrogen content prediction in winter wheat.

However, it is undeniable that this study has limitations in exploring the transferability of the model based only on data from different irrigation treatments at the same sampling sites. Regarding the transferability of the model, the applicability between different regions, species, and years will be the next direction of our research.

Acknowledgements

This research was funded by the Henan Province Key R&D and Promotion Projects (No. 232102111030), and the Science and Technology Innovation Leading Talent Cultivation Program of the Institute of Agricultural Economics and Information, Henan Academy of Agricultural Sciences (No. 2022KJCX01). The research was also funded by the Independent innovation projects of Henan Academy of Agricultural Sciences (No. 2023ZC062).

Conflicts of interest

The authors declare no conflict of interest.

Author details

Yan Guo[1,2], Jia He[1], Jingyi Huang[3], Xiuzhong Yang[1,2], Zhou Shi[4], Laigang Wang[1,5]* and Guoqing Zheng[1,2]*

1 Institute of Agricultural Economics and Information, Henan Academy of Agricultural Sciences/Key Laboratory of Huang-Huai-Hai Smart Agricultural Technology, Ministry of Agriculture and Rural Affairs, P.R. China, Zhengzhou, China

2 Henan Engineering Research Center of Crop Planting Monitoring and Warning, Zhengzhou, China

3 Department of Soil Science, University of Wisconsin-Madison, Madison, USA

4 Institute of Applied Remote Sensing and Information Technology, College of Environmental and Resource Sciences, Zhejiang University, Hangzhou China

5 International School, Huanghe Science and Technology College, Zhengzhou, China

*Address all correspondence to: wlaigang@sina.com and zgqzx@hnagri.org.cn

References

[1] World agricultural production. In: Circular Series WAP. Washington: United States Department of Agriculture (USDA); 2022. pp. 27-42. Available from: https://apps.fas.usda.gov/psdonline/circulars/production.pdf

[2] Shiferaw B, Smale M, Braun H, Duveiller E, Reynolds MP, Muricho G. Crops that feed the world 10. Past successes and future challenges to the role played by wheat in global food security. Food Security. 2013;**5**:291-317. DOI: 10.1007/s12571-013-0263-y

[3] Perry EM, Fitzgerald GJ, Nuttall JG, O'Leary G, Schulthess U, Whitlock A. Rapid estimation of canopy nitrogen of cereal crops at paddock scale using a canopy chlorophyll content index. Field Crops Research. 2023;**134**:158-164. DOI: 10.1016/j.fcr.2012.06.003

[4] Gu B, Zhang X, Lam SK, Yu Y, van Grinsven HJ, Zhang S, et al. Cost-effective mitigation of nitrogen pollution from global croplands. Nature. 2023;**613**:77-84. DOI: 10.1038/s41586-022-05481-8

[5] Raj R, Walker JP, Pingale R, Banoth BN, Jagarlapudi A. Leaf nitrogen content estimation using top-of-canopy airborne hyperspectral data. International Journal of Applied Earth Observation and Geoinformation. 2021;**104**:102584. DOI: 10.1016/j.jag.2021.102584

[6] De Silva AL, Trueman SJ, Kämper W, Wallace HM, Nichols JT, Hosseini BS. Hyperspectral imaging of adaxial and abaxial leaf surfaces as a predictor of macadamia crop nutrition. Plants. 2023;**12**:558. DOI: 10.3390/plants12030558

[7] Jiang JL, Zhu J, Wang X, Cheng T, Tian YC, Zhu Y, et al. Estimating the leaf nitrogen content with a new feature extracted from the ultra-high spectral and spatial resolution images in wheat. Remote Sensing. 2021;**13**:739. DOI: 10.3390/rs13040739

[8] Sabzi S, Pourdarbani R, Rohban MH, García-Mateos G, Arribas JI. Estimation of nitrogen content in cucumber plant (Cucumis sativus L.) leaves using hyperspectral imaging data with neural network and partial least squares regressions. Chemometrics and Intelligent Laboratory Systems. 2021;**217**:104404. DOI: 10.1016/j.chemolab.2021.104404

[9] Guo Y, He J, Huang J, Jing Y, Xu S, Wang L, et al. Effects of the spatial resolution of UAV images on the prediction and transferability of nitrogen content model for winter wheat. Drones. 2022;**6**:299. DOI: 10.3390/drones6100299

[10] Ruan G, Schmidhalter U, Yuan F, Cammarano D, Liu X, Tian Y, et al. Exploring the transferability of wheat nitrogen status estimation with multisource data and evolutionary algorithm-deep learning (EA-DL) framework. European Journal of Agronomy. 2023;**143**:126727. DOI: 10.1016/j.eja.2022.126727

[11] Yang BH, Chen JL, Chen LH, Cao WX, Yao X, Zhu Y. Estimation model of wheat canopy nitrogen content based on sensitive bands. Transactions of the Chinese Society of Agricultural Engineering. 2015;**31**:176-182. DOI: 10.11975/j.issn.1002-6819.2015.22.024 (in Chinese)

[12] Zhang XY, Zhang LF, Zhang X, Wang SD, Tian JG, Zhai YG. Sensitivity of different spectral vegetation index for estimating winter wheat leaf nitrogen. Scientia Agricultura Sinica. 2017;**50**:474-485. DOI: 10.3864/j.issn.0578-1752.2017.03.006 (in Chinese)

[13] Li Z, Li Z, Fairbairn D, Li N, Xu B, Feng H, et al. Multi-LUTs method for canopy nitrogen density estimation in winter wheat by field and UAV hyperspectral. Computers and Electronics in Agriculture. 2019;**162**:174-182. DOI: 10.1016/j.compag.2019.04.005

[14] Jia D, Cheng PF. Effect of low altitude UAV image resolution on inversion of winter wheat nitrogen concentration. Transactions of the Chinese Society of Agricultural Machinery. 2020;**51**:164-169. DOI: 10.6041/j.issn.1000-1298.2020.07.019 (in Chinese)

[15] Fu Y, Yang G, Song X, Li Z, Xu X, Feng H, et al. Improved estimation of winter wheat aboveground biomass using multiscale textures extracted from UAV-based digital images and hyperspectral feature analysis. Remote Sensing. 2021;**13**:581. DOI: 10.3390/rs13040581

[16] Gandhimathi Alias Usha S, Vasuki S. Significance of texture features in the segmentation of remotely sensed images. Optik. 2022;**249**:168241. DOI: 10.1016/j.ijleo.2021.168241

[17] Zheng H, Ma J, Zhou M, Li D, Yao X, Cao W, et al. Enhancing the nitrogen signals of rice canopies across critical growth stages through the integration of textural and spectral information from unmanned aerial vehicle (UAV) multispectral imagery. Remote Sensing. 2020;**12**:957. DOI: 10.3390/rs12060957

[18] Wang X, Miao Y, Dong R, Zha H, Xia T, Chen Z, et al. Machine learning-based in-season nitrogen status diagnosis and side-dress nitrogen recommendation for corn. European Journal of Agronomy. 2021;**123**:126193. DOI: 10.1016/j.eja.2020.126193

[19] Dehghan-Shoar MH, Orsi AA, Pullanagari R, Yule I. A hybrid model to predict nitrogen concentration in heterogeneous grassland using field spectroscopy. Remote Sensing of Environment. 2023;**285**:113385. DOI: 10.1016/j.rse.2022.113385

[20] Chlingaryan A, Sukkarieh S, Whelan B. Machine learning approaches for crop yield prediction and nitrogen status estimation in precision agriculture: A review. Computers and Electronics in Agriculture. 2018;**151**:61-69. DOI: 10.1016/j.compag.2018.05.012

[21] Berger K, Verrelst J, Féret J, Hank TB, Wocher M, Mauser W, et al. Retrieval of aboveground crop nitrogen content with a hybrid machine learning method. International Journal of Applied Earth Observation and Geoinformation. 2020;**92**:102174. DOI: 10.1016/j.jag.2020.102174

[22] Qiu Z, Ma F, Li Z, Xu X, Ge H, Du C. Estimation of nitrogen nutrition index in rice from UAV RGB images coupled with machine learning algorithms. Computers and Electronics in Agriculture. 2021;**189**:106421. DOI: 10.1016/j.compag.2021.106421

[23] Zhang SH, Duan JZ, He L, Jing YH, Schulthess UC, Lashkari A, et al. Wheat yield estimation from UAV platform based on multi-modal remote sensing data fusion. Scientia Agricultura Sinica. 2022;**48**:1746-1760. DOI: 10.3724/SP.J.1006.2022.11053 (In Chinese)

[24] Gamon JA, Peñuelas J, Field CB. A narrow-waveband spectral index that tracks diurnal changes in photosynthetic efficiency. Remote Sensing of Environment. 1992;**41**:35-44. DOI: 10.1016/0034-4257(92)90059-S

[25] Xiao X, He L, Salas WA, Li C, Moore B, Zhao R, et al. Quantitative relationships between field-measured leaf area index and vegetation index derived from vegetation images for

paddy rice fields. International Journal of Remote Sensing. 2002;**23**:3595-3604. DOI: 10.1080/01431160110115799

[26] Zhou Y, Lao C, Yang Y, Zhang Z, Chen H, Chen Y, et al. Diagnosis of winter-wheat water stress based on UAV-borne multispectral image texture and vegetation indices. Agricultural Water Management. 2021;**256**:107076. DOI: 10.1016/j.agwat.2021.107076

[27] Soltanikazemi M, Minaei S, Shafizadeh-Moghadam H, Mahdavian AR. Field-scale estimation of sugarcane leaf nitrogen content using vegetation indices and spectral bands of Sentinel-2: Application of random forest and support vector regression. Omputers and electronics. Agriculture. 2022;**200**:107130. DOI: 10.1016/j.compag.2022.107130

[28] Niu YX, Zhang LY, Han WT, Shao GM. Fractional vegetation cover extraction method of winter wheat based on UAV remote sensing and vegetation index. Transactions of the Chinese Society of Agricultural Machinery. 2018;**49**:212-221. DOI: 10. 6041/j. issn.1000-1298.2018.04.024 (in Chinese)

[29] Jia M, Colombo R, Rossini M, Celesti M, Zhu J, Cogliati S, et al. Estimation of leaf nitrogen content and photosynthetic nitrogen use efficiency in wheat using sun-induced chlorophyll fluorescence at the leaf and canopy scales. European Journal of Agronomy. 2021;**122**:126192. DOI: 10.1016/j. eja.2020.126192

[30] Gitelson AA, Kaufman YJ, Merzlyak MN. Use of a green channel in remote sensing of global vegetation from EOS-MODIS. Remote Sensing of Environment. 1996;**58**:289-298. DOI: 10.1016/S0034-4257(96)00072-7

[31] Gilabert MA, González-Piqueras J, García-Haro FJ, Meliá J. A generalized soil-adjusted vegetation index. Remote Sensing of Environment. 2002;**82**:303-310. DOI: 10.1016/ S0034-4257(02)00048-2

[32] Rouse JW. Monitoring the vernal advancement of retrogradation (green wave effect) of natural vegetation. NASA/GSFC Type III. final Report. greenbelt Md. 1974. p. 371

[33] Chen JM. Evaluation of vegetation indices and a modified simple ratio for boreal applications. Canadian Journal of Remote Sensing. 1996;**22**:229-242. DOI: 10.1080/07038992.1996.10855178

[34] Imran AB, Khan K, Ali N, Ahmad N, Ali AM, Shah K. Narrow band based and broadband derived vegetation indices using Sentinel-2 imagery to estimate vegetation biomass. Global Journal of Environmental Science and Management. 2020;**6**:97-108. DOI: 10.22034/gjesm.2020.01.08

[35] Rondeaux G, Baret F, Steven M. Optimization of soil-adjusted vegetation indices. Remote Sensing of Environment. 1996;**55**:95-107. DOI: 10.1016/0034-4257(95)00186-7

[36] Gitelson AA, Viña A, Ciganda VS, Rundquist D, Arkebauer TJ. Remote estimation of canopy chlorophyll content in crops. Geophysical Research Letters. 2005;**32**:1-4. DOI: 10.1029/2005GL022688

[37] Huete A, Justice C, Leeuwen WV. MODIS Vegetation Index (MOD13) Algorithm Theoretical Basis Document. US: NASA Goddard Space Flight Center; 1999

[38] Qian B, Ye H, Huang W, Xie Q, Pan Y, Xing N, et al. A sentinel-2-based triangular vegetation index for chlorophyll content estimation. Agricultural and Forest Meteorology. 2022;**322**:109000. DOI: 10.1016/j. agrformet.2022.109000

[39] Kaufman YJ, Tanré D. Atmospherically resistant vegetation index (ARVI) for EOS-MODIS. IEEE Transactions on Geoscience and Remote Sensing. 1992;**30**:261-270. DOI: 10.1109/36.134076

[40] Lamm RD, Slaughter DC, Giles DK. Precision weed control system for cotton. Transactions of the ASABE. 2002;**45**:231-238. DOI: 10.13031/2013.7861

[41] Birth GS, Mcvey GR. Measuring the color of growing turf with a reflectance spectrophotometer. Agronomy Journal. 1968;**60**:640-643. DOI: 10.2134/AGRONJ 1968.00021962006000060016X

[42] Haboudane D, Miller JR, Pattey E, Zarco-Tejada PJ, Strachan IB. Hyperspectral vegetation indices and novel algorithms for predicting green LAI of crop canopies: Modeling and validation in the context of precision agriculture. Remote Sensing of Environment. 2004;**90**:337-352. DOI: 10.1016/J.RSE.2003.12.013

[43] Huete AR, Hua G, Qi J, Chehbouni AG, Leeuwen WJ. Normalization of multidirectional red and NIR reflectances with the SAVI. Remote Sensing of Environment. 1992;**41**:143-154. DOI: 10.1016/0034-4257(92)90074-T

[44] Jordan CF. Derivation of leaf-area index from quality of light on the forest floor. Ecology. 1969;**50**:663-666. DOI: 10.2307/1936256

[45] Reyniers M, Walvoort DJJ, De Baardemaaker J. A linear model to predict with a multi-spectral radiometer the amount of nitrogen in winter wheat. International Journal of Remote Sensing. 2006;**27**:4159-4179. DOI: 10.1080/01431160600791650

[46] Yadav V, Bharadi VA, Yadav SK. Texture feature extraction using hybrid wavelet type I & II for finger knuckle prints for multi-algorithmic feature fusion. Procedia Computer Science. 2016;**79**:359-366. DOI: 10.1016/J.PROCS.2016.03.047

[47] Yang Y, Chen J. Comprehensive analysis of water carrying capacity based on wireless sensor network and image texture of feature extraction. Alexandria Engineering Journal. 2022;**61**:2877-2886. DOI: 10.1016/j.aej.2021.08.018

[48] Du B, Lund PD, Wang J, Kolhe M, Hu E. Comparative study of modelling the thermal efficiency of a novel straight through evacuated tube collector with MLR, SVR, BP and RBF methods. Sustainable Energy Technologies and Assessments. 2021;**44**:101029. DOI: 10.1016/j.seta.2021.101029

[49] Tang X, Liu H, Feng D, Zhang W, Chang J, Li L, et al. Prediction of field winter wheat yield using fewer parameters at middle growth stage by linear regression and the BP neural network method. European Journal of Agronomy. 2022;**121**:126621. DOI: 10.1016/j.eja.2022.126621

[50] Zhou ZH. Machine Learning. Beijing: Tsinghua University Press; 2016 (in Chinese)

[51] Jeung M, Baek S, Beom J, Cho KH, Her Y, Yoon KS. Evaluation of random forest and regression tree methods for estimation of mass first flush ratio in urban catchments. Journal of Hydrology. 2019;**575**:1099-1110. DOI: 10.1016/j.jhydrol.2019.05.079

[52] Fernández-Habas J, Cañada MC, Moreno AM, Leal-Murillo JR, González-Dugo MP, Oar BA, et al. Estimating pasture quality of Mediterranean grasslands using hyperspectral narrow bands from field spectroscopy by random Forest and PLS regressions. Computers and Electronics in Agriculture. 2022;**192**:106614. DOI: 10.1016/j.compag.2021.106614

[53] Liepe MU, Belomestnykh S. RF parameter and field stability requirements for the Cornell ERL prototype. In: Proceedings of the 2003 Particle Accelerator Conference. May 12-17, 2003; Portland, OR. New York: IEEE; Vol. 2. 2003. pp. 1329-1331. DOI: 10.1109/PAC.2003.1289695

[54] Barrow DK, Crone SF. A comparison of Adaboost algorithms for time series forecast combination. International Journal of Forecasting. 2016;**32**:1103-1119. DOI: 10.1016/j.ijforecast.2016.01.006

[55] Wang LL, Guo YL, Fan MH, Li X. Wind speed prediction using measurements from neighboring locations and combining the extreme learning machine and the Adaboost algorithm. Energy Reports. 2022;**8**:1508-1518. DOI: 10.1016/j.egyr.2021.12.062

[56] Zhou Y, Lu Z, Chen GK. Adaboost-based ensemble of polynomial chaos expansion with adaptive sampling. Computer Methods in Applied Mechanics and Engineering. 2022;**388**:114238. DOI: 10.1016/j.cma.2021.114238

[57] Panahi M, Sadhasivam N, Pourghasemi HR, Rezaie F, Lee S. Spatial prediction of groundwater potential mapping based on convolutional neural network (CNN) and support vector regression (SVR). Journal of Hydrology. 2020;**588**:125033. DOI: 10.1016/j.jhydrol.2020.125033

[58] Alnuwaiser MA, Javed MF, Khan MI, Ahmed MW, Galal AM. Support vector regression and ANN approach for predicting the ground water quality. Journal of the Indian Chemical Society. 2022;**99**:10538. DOI: 10.1016/j.jics.2022.100538

[59] Li Y, Sun H, Yan W, Zhang X. Multi-output parameter-insensitive kernel twin SVR model. Neural networks: the Official Journal of the International Neural Network Society. 2020;**121**:276-293. DOI: 10.1016/j.neunet.2019.09.022

[60] Yi Y, Wang L, Chen Z. Adaptive global kernel interval SVR-based machine learning for accelerated dielectric constant prediction of polymer-based dielectric energy storage. Renewable Energy. 2021;**176**:81-88. DOI: 10.1016/J.RENENE.2021.05.045

[61] Bellis ES, Hashem AA, Causey JL, Runkle B, Moreno-García B, Burns BW, et al. Detecting intra-field variation in rice yield with unmanned aerial vehicle imagery and deep learning. Frontiers in Plant Science. 2022;**13**:716506. DOI: 10.3389/fpls.2022.716506

[62] Lin LX, Liu XX. Mixture-based weight learning improves the random forest method for hyperspectral estimation of soil total nitrogen. Computers and Electronics in Agriculture. 2022;**192**:106634. DOI: 10.1016/j.compag.2021.106634

[63] Ng A. Nuts and bolts of building applications using deep learning. Available from: https://neurips.cc/Conferences/2016

[64] Linardatos P, Papastefanopoulos V, Kotsiantis SB. Explainable AI: A review of machine learning interpretability methods. Entropy. 2021;**23**:18. DOI: 10.3390/e23010018

[65] Sarker IH. Machine learning: Algorithms, real-world applications and research directions. SN Computer Science. 2021;**2**:160. DOI: 10.1007/s42979-021-00592-x

[66] Wang J, Shi T, Yu D, Teng D, Ge X, Zhang Z, et al. Ensemble machine-learning-based framework for estimating total nitrogen concentration in water using drone-borne hyperspectral imagery of emergent plants: A case study in an arid oasis. NW China. Environmental Pollution. 2020;**266**:115412. DOI: 10.1016/j.envpol.2020.115412

Chapter 13

Using Unmanned Aerial Systems and Deep Learning for Agriculture Mapping in Dubai Emirate

Lala El Hoummaidi, Abdelkader Larabi and Khan Alam

Abstract

Dubai's 'Sustainable Future' vision prioritizes Sustainable Agriculture as a key pillar of its 'Food Security Strategies'. To boost productivity and efficiency, Dubai Emirate has adopted advanced technologies. Accurate land monitoring is crucial for effective food security control and support measures. However, traditional methods relying on costly and time-consuming field surveys conducted by experts are limited in scope. To address this, affordable and efficient agriculture mapping relies on remote sensing through drone surveys. Dubai Municipality utilizes Unmanned Aerial Vehicles (UAVs) to map farming areas across the Emirate, identify cultivable lands, and establish a precise agriculture database. A study conducted over 6 months used Trimble UX5 (HP) drones for high-resolution imaging in 12 Dubai communities. It employed novel object detection methods and geospatial analysis. Deep learning models achieved 85.4% accuracy in vegetation cover and F1-scores of 96.03% and 94.54% for date palms and GHAF trees, respectively, compared to ground truth data. This research highlights the potential of UAVs and deep learning algorithms for large-scale sustainable agricultural mapping. By providing specialists with an integrated solution to measure and assess live green vegetation cover derived from processed images, it contributes to the advancement of sustainable agriculture practices.

Keywords: precision agriculture, multispectral imaging, UAV, remote sensing, machine learning, deep learning

1. Introduction

Undoubtedly, agriculture plays a pivotal role in ensuring the sustainability of economies [1, 2]. Its significance, however, may vary across different countries [3–6]. While agriculture was traditionally limited to food and crop production, it has now expanded in numerous countries to encompass processing, marketing, and distribution of agricultural products. Agricultural activities not only serve as a primary source of livelihood and contribute to GDP growth [7], but also drive national trade, reduce unemployment, provide raw materials for other industries, and contribute to overall economic development [8–10].

Remote sensing techniques, including soil property mapping, crop type classification, crop water stress detection, disease monitoring, and crop yield mapping, have gained widespread adoption in both governmental and private sectors [11, 12]. Leveraging sensors and geospatial analysis tools, remote sensing brings together data from multiple sources to support decision-making in agriculture. Unmanned Aerial Systems (UAS) or drones, with their flexible spatial and spectral resolution, have become a preferred platform for collecting such data [13]. Additionally, remote sensing-based land cover classification has found applications in change detection monitoring, agricultural management, green vegetation classification, biodiversity conservation, land use planning, and urban planning [14, 15]. Vegetation detection stands out as a significant application of land cover classification. Consequently, researchers and experts have explored various methods, including digital photo interpretation, supervised and unsupervised classification, classification and regression trees (CART), and Deep Learning Object Detection [16–20].

Deep learning techniques have gained prominence in land cover classification since 2012, furthermore, due to remarkable progress in computer vision applications like image classification, object detection, tracking, and semantic segmentation, researchers have been able to explore various methodologies. Snehal et al. [21] utilized convolutional networks to tackle the challenge of multispectral image classification. In a separate study, Zhang et al. [22] conducted an in-depth analysis of different techniques for object detection in land cover classification, particularly focusing on the utilization of high-resolution multispectral imagery. Their study compared deep learning models with traditional methods, concluding that deep learning-based approaches leveraging both spatial and spectral information outperformed conventional pixel-based methods. Other studies have also demonstrated the potential of artificial intelligence and deep learning methods for land cover classification and vegetation detection [23–25].

Amidst all-time low commodity prices and increasing pressure for enhanced product quality, the modern farming industry faces challenges that necessitate improved resource management. Dubai Emirate shares this need, as it aims to leverage its re-export hub and global gateway status in the fresh food sales sector. Notably, organic farms in Dubai witnessed a 53% increase in 2019, with production rising by 89% from 1240 tons to 2356 tons [26]. To support an efficient workflow for assessing crop health, making informed decisions, and mitigating losses due to disease outbreaks or extreme weather events, Dubai Municipality has initiated projects utilizing drones and connected analytics for surveying and mapping agricultural areas. High-resolution multispectral drones enable growers, service providers, and researchers to efficiently scout crops, identify stress, track plant growth, and access real-time quality data, ultimately reducing costs and improving yields. Moreover, multispectral data reveals field variability invisible to the naked eye, aiding in early disease detection and response.

This chapter focuses on evaluating the suitability of UAS-based remote sensing, using a novel object-based vegetation detection method that combines NDVI and deep learning techniques, for monitoring crops in Dubai. The contributions of this study include the introduction of a superior object-based vegetation and tree detection method using NDVI and deep learning, highlighting the potential use of NDVI imagery as an alternative to standard RGB images, and discussing the reasons behind the superior performance of our deep learning model compared to other methods, along with potential strategies for its further application.

2. Materials and methods

2.1 Study area

Dubai Emirate, the second largest among the seven emirates comprising the United Arab Emirates, is strategically located along the southeastern coast of the Arabian Gulf, spanning from coordinates 55°18′14.8188″ East to 25°16′ 17.4564”North. Encompassing a combined expanse of 3900 square kilometers, the emirate extends for about 72 kilometers along the coastline of the Arabian Gulf. In terms of its geographical location, Dubai is positioned adjacent to Sharjah in the northeast, the capital city of Abu Dhabi to the south, and the Sultanate of Oman to the southeast. The administrative boundaries of the Dubai Emirate define its territorial extent, depicted in **Figure 1**, delineate its territorial extent, accounting for approximately 5% of the total area of the United Arab Emirates.

The diverse landscape of Dubai encompasses shallow shores, sandy deserts, and coral reefs. With over 300 species of fish inhabiting its waters, the rich marine life of Dubai has served as a significant source of income for its residents for thousands of years. Date palms dominate much of Dubai's cultivated land, primarily found in the arc of small oases that make up the Hatta Area. Dubai Municipality supports farmers through various incentives, such as a 50% subsidy on fertilizers, seeds, and pesticides. Additionally, it offers loans for machinery and provides technical assistance [27, 28].

According to the data presented in **Table 1**, vegetable cultivation accounts for 13% of the total cultivated land in Dubai Emirate, while fruit crops occupy 32%, feed crops cover 14%, and the remaining 43% is allocated for various other uses. Notably, the

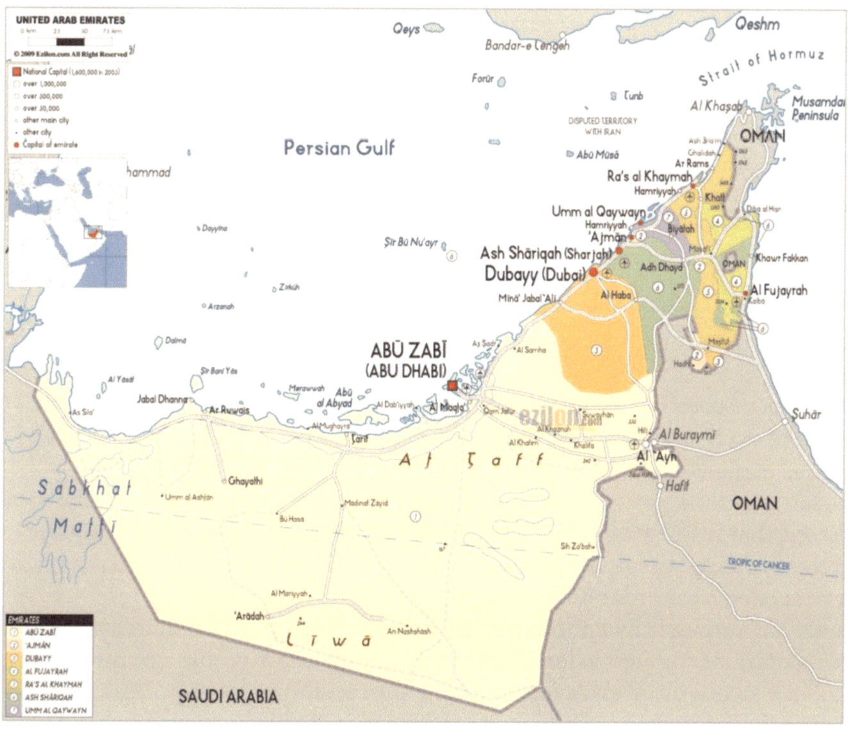

Figure 1.
Dubai emirate and major cities within the United Arab Emirates.

Years	Forest trees (dūnum)	Vegetables (dūnum)	Fruit trees (dūnum)	Feed crops (dūnum)	Temporary fallow (dūnum)	Other lands (dūnum)	Total (dūnum)
2017	760	1765	15,810	3177	1725	21,645	44,882
2018	784	1660	15,903	3218	1664	18,464	41,693
2019	784	7529	20,409	8108	1917	23,168	61,914
Area %	1%	13%	32%	14%	3%	37%	100%

Table 1.
Land use distribution in Dubai, area in donums (dūnum) - 2017–2019.

Crop	Value	Average of production	Quantity	Area
	(000 AED)	(Tons/Dūnum)	(Tons)	(Dūnum)
Tomatoes	18,030.9	6.2	6297.9	1010.8
Cucumber	8166.9	9.4	2946.7	313.9
Pepper	1637.1	4.9	403.3	82.5
Squash	8517.3	2.3	2793.9	1219.8
Eggplants	4575.6	3.3	2333.8	716.6
Cauliflower	7216.0	3.5	2590.3	740.1
Cabbage	7488.8	6.0	4082.0	680.3
Water melon	474.9	3.0	391.9	130.3
Leafy vegetables	1740.7	1.3	579.9	447.1
Other	10,567.6	2.3	4970.5	2187.3
Total	68,415.8	3.6	27,390.3	7528.8

Table 2.
Vegetables by crop in Dubai – 2019.

region of Hatta demonstrates high productivity due to its access to underground water sources from the nearby mountains of Oman, benefiting from abundant rainfall. In this region, the main crops cultivated include tomatoes, melons, and dates. Despite the challenges posed by the desert environment, vegetable production in Dubai has successfully overcome obstacles, resulting in a noteworthy production of over 27 tons in 2019, as depicted in **Table 2**.

Tomatoes, cabbage, eggplant, squash, and cauliflower serve as the primary vegetable crops that fulfill a significant portion of the Emirate's demand during the respective growing season. Furthermore, citrus fruits and mangoes are the main fruit crops cultivated in addition to dates [29].

Dubai has successfully addressed significant challenges, including harsh environmental conditions, limited water resources, and soil salinity, through innovative solutions. These solutions often involve tapping into underground aquifers or accessing water supplies from the mountains. As indicated in **Table 3**, agriculture in Dubai is practiced on approximately 8000 dunums of cultivable land, with a significant portion dedicated to Couch grass and Alfalfa.

Crop	Value	Average of production	Quantity	Area
	(000 AED)	(Tons/Dūnum)	(Tons)	(Dūnum)
Alfalfa	20,056.8	6.0	12,535.5	2089.2
Rhode grass	32,362.3	6.0	21,574.9	3595.8
Sorghum	11,953.0	6.0	7470.6	1245.1
Maize	2059.4	2.8	2059.4	748.9
Other	3343.1	6.0	2571.6	428.6
Total	**69,774.6**	**5.7**	**46,212.0**	**8107.6**

Table 3.
Total and area of field crops by crop in Dubai – 2019.

2.2 Overall study workflow

The workflow, illustrated in **Figure 2**, consists of multiple steps: acquiring multispectral imagery, labeling, dataset preparation, model training, object detection, result analysis, and field validation. The ArcGIS API for Python was used to export training data and train deep learning models within the study area. PyTorch and fast.ai libraries were utilized for data preparation, augmentation, and model training. The object detection models employed the PASCAL_VOC_rectangles format for training samples. Data preparation involved organizing and formatting training labels, splitting data, applying augmentation, and creating data structures. ArcGIS Pro facilitated these tasks, allowing direct reading of training samples and creating a suitable DataBunch with specified parameters.

2.3 Drone data

Figure 3 depicts the various elements comprising the Trimble UX5 HP Unmanned Aircraft System (UAS), which serves as the primary tool for field data capture. This UAS device is user-friendly, fully automated, and capable of capturing high-resolution aerial photography with resolutions as fine as 1 cm. It offers an intuitive workflow that facilitates the efficient creation of top-quality ortho-mosaics and 3 Advanced three-dimensional (3D) models have been developed for a variety of applications within the realm of agriculture. These applications encompass agriculture mapping, field leveling, progress monitoring, and asset mapping [30].

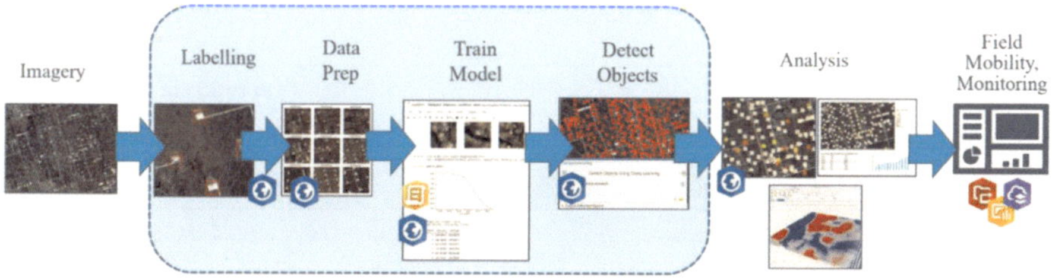

Figure 2.
Complete workflow illustration, encompassing the transformation of raw imagery into structured information about vegetation cover feature layers.

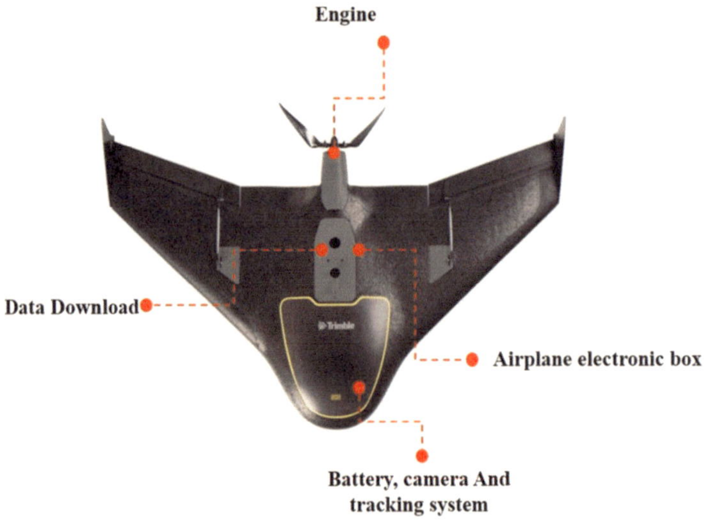

Figure 3.
Utilization of Trimble drones for data collection in Dubai emirate.

In Dubai Emirate, the team operating in the field utilizes the capabilities of high-resolution multispectral drone imagery to effectively capture valuable data for their operations, enabling the generation of Normalized Difference Vegetation Index (NDVI) maps. These maps are instrumental in distinguishing between soil, forests, and grass, as well as identifying different crop stages and detecting plants under stress. Significant research findings have firmly established robust associations between crop yield and specifically measured NDVI (Normalized Difference Vegetation Index) data at distinct stages of crop growth. Consequently, monitoring crop growth at key stages enables precise crop yield estimation and early identification and resolution of issues [31, 32]. For specific details regarding the acquisition performance levels of the Trimble UX5 Drone, please refer to **Table 4**.

Dubai Municipality employs a drone equipped with state-of-the-art photogrammetric and navigation equipment, granting it a ground resolution capacity of up to 3 centimeters. Its programming allows for the detection of crucial details such as NDVI, water stress, and specific nutrient deficiencies in crops. The Geographic Information Systems Centre (GISC) in Dubai Emirate seamlessly integrates drone-based mapping efforts into disaster risk reduction and management (DRRM) and climate change adaptation (CCA) strategies. The Trimble UX5 HP drone, equipped with a modified color-infrared (CIR) Sony NEX5R camera and a 16 mm lens, was employed during field surveys. Each flight day involved gathering around 100 ground-based

Resolution (GSD)	1 cm to 25 cm (4 to 99 in)
Height above take-off location (AGL)	75 m to 750 m (246 to 2460 feet)
Absolute accuracy XY/Z (no ground control points)	down to 2–5 cm
Relative Ortho-mosaic/3D model accuracy	(1–2x/1–5x GSD)
Resolution (GSD)	1 cm to 25 cm (4 to 99 in)

Table 4.
Acquisition performance parameters for Trimble UX5 drone.

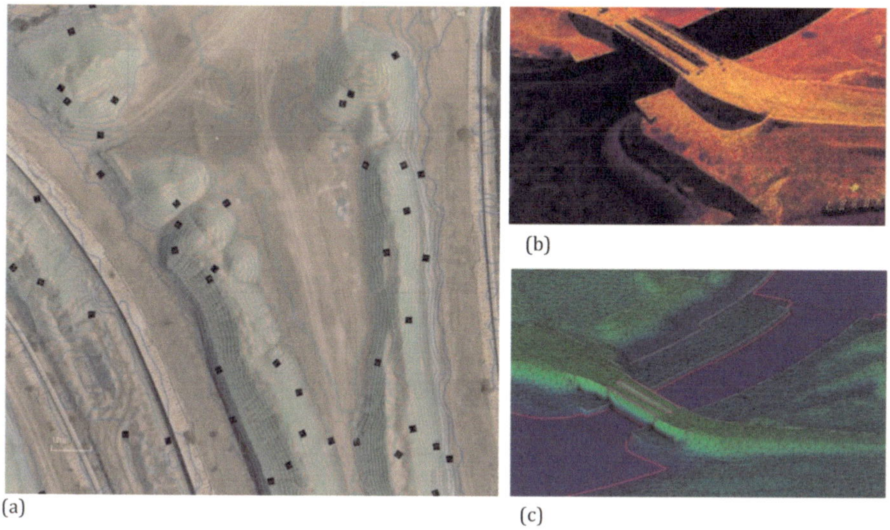

Figure 4.
Outputs of drone mapping: (a) rectified image with contours, (b) 3D surface created from point clouds, (c) 3D surface constructed from contour lines.

normalized difference vegetation index (NDVI) measurements using the Trimble Green Seeker Handheld device, consistently maintaining a height of 80 cm above the target. The georeferencing process ensured a geospatial accuracy of 2 cm using a Trimble R8RTK GNSS system [33]. **Figure 4** showcases sample outputs from the drone, including an ortho-rectified image with elevation contours, a three-dimensional surface generated from collected point clouds, and a three-dimensional surface created using processed contour lines.

Aerial surveys are crucial, operating in real airspace alongside other aircraft. The process involves creating a flight plan, estimating Ground Control Points (GCP), and conducting a risk assessment. **Figure 5** illustrates capturing aerial data with multi-spectral cameras and LIDAR using Trimble UX5. Surveys cover areas like Hatta in 4 days, while processing takes 1 day for DEM and Ortho-Photo generation.

In this study, a total area of 770 square kilometers across 12 areas was comprehensively surveyed using five teams over the course of 139 days. The subsequent data processing phase took around 78 days to complete. Some areas, like AL WOHOOCH and SAIH SHUAIB, were easier to survey due to their uniform flat sand dune surfaces. Please refer to **Table 5** for a detailed breakdown of flying time, processing time, number of flights, and repetitions for each community in 2019.

The modified Sony NEX5R camera captures 3-band R-G-NIR imagery, with blue-filtered pixels receiving NIR and green/red filters capturing visible light along with red edge and NIR wavelengths. It stored both 14-bit per band linear lossy compressed RAW files (35 MB each) and 8-bit per band gamma-compressed JPEG files (15 MB each) [34]. See **Figure 6** for the Trimble UX5 device's true color (RGB) spectral responses.

2.4 Processing and analysis

The methodology developed for processing in this study involved four main steps: photogrammetric pre-processing, deep learning-based object detection, data analysis,

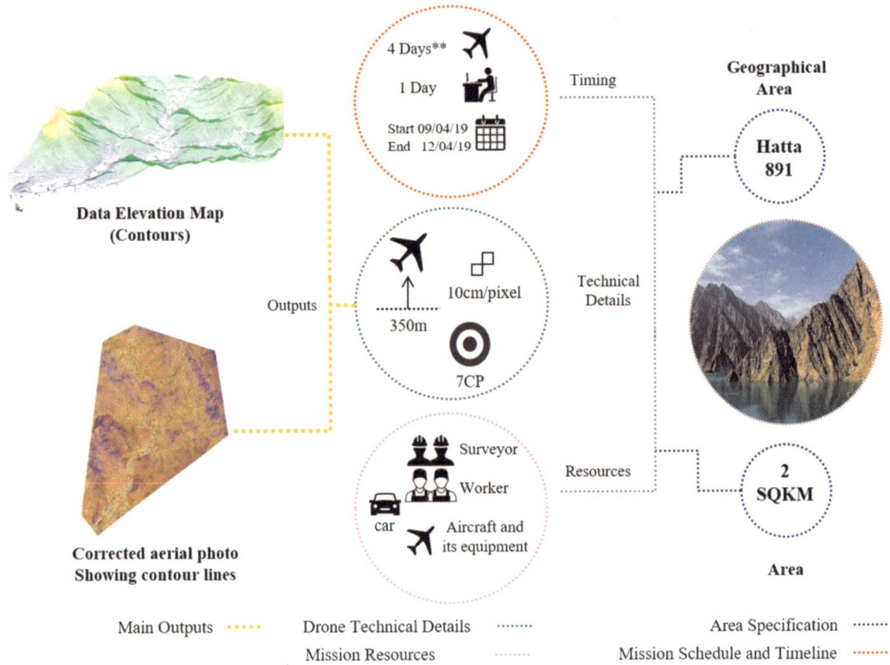

Figure 5.
Process of drone mapping field Mission in Hatta region.

Community	SQKM	Flying time in days	Processing time in days	Number of flights	Repetitions per year (2019)
Saih Shuaib	41.61	5	4	5	1
Hadaeq Sheikh Mohammed bin Rashid	38.68	19	3	38	2
Aleyas	10.52	5	8	5	1
Al Kheeran	7.33	4	6	12	3
Al Lesaily	112.69	13	10	13	1
Margham	152.59	25	12	25	1
Al Wohoosh	26.51	3	2	3	1
Al Maha	41.73	21	4	42	2
Remah	82.87	5	7	5	1
Grayteesah	91.83	12	8	12	1
Al Fagaa	140.53	15	13	15	1
Hessyan	23.85	12	1	12	1
Total	**770.75**	**139**	**78**	**187**	

Table 5.
Details of unmanned aerial system (UAS) missions for the designated study areas.

and result evaluation. The initial step focused on pre-processing the imagery obtained from the Unmanned Aerial System (UAS) using digital photogrammetry techniques. Next, deep learning algorithms were carefully selected and implemented to detect

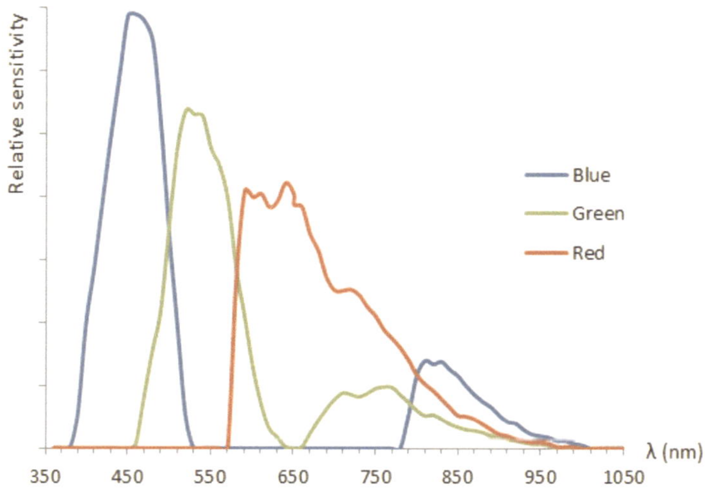

Figure 6.
Spectral response of the Trimble UX5 HP Sony NEX-5 N.

vegetation cover, identify diseases, and perform template matching for segmenting the main crop area and detecting individual crops on the orthophoto mosaic. Subsequently, advanced geoprocessing tools were employed for data analysis. Finally, the detection accuracy threshold was established, and a comprehensive comparison was conducted between crop volume, estimated crop pests, and field samples [35].

2.4.1 Pre-processing

Data layers, including ground measurements, were aligned and registered using ArcGIS software. UAS-based NDVI values and spectral profiles were evaluated for multitemporal monitoring. Point features were interpolated using natural neighbor interpolation. A digital terrain model was generated for crop height calculation.

The use of topographic maps is crucial for agriculture and vegetation mapping [36], as certain species thrive at specific elevation levels. Therefore, data captured using drones and aligned through aerial triangulation, ortho-rectification, and georeferencing using ground control point (GCP) information is essential. **Figure 7** illustrates the final Digital Elevation Model (DEM) and contour lines generated for the Hatta Region using ortho-rectified drone imagery.

2.4.2 Object detection algorithms (deep learning)

Deep learning models were evaluated in ArcGIS for classifying tree points in geospatial datasets. Two models, "Tree Point Classification" and "Landcover Classification," were modified for the arid environment of Dubai using Python scripts. These models successfully detected vegetation cover and individual trees with TensorFlow. The workflow, shown in **Figure 8**, involved data preparation, training with ArcGIS.learn module, and deployment as deep learning packages. All algorithms were implemented in Python, and experiments were conducted on a high-performance workstation.

The most challenging aspect of the work involved data preparation, training sample creation, and model training to extract features from the imagery. These steps have been completed, and a trained model is now utilized to detect various types of

Figure 7.
Generated contour map for Hatta region.

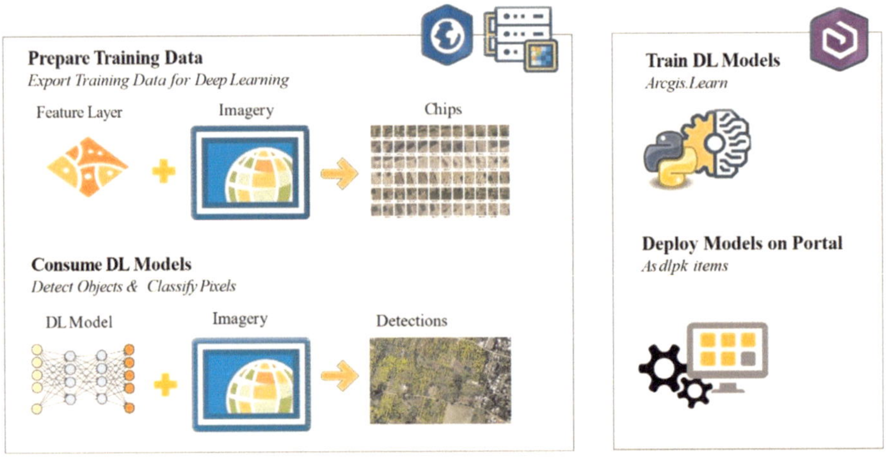

Figure 8.
Usage workflow of the General Deep Learning Packages (DLPKs).

crops in the processed drone imagery. Achieving optimal results in object detection requires thorough testing and adjustment of parameters. The model's performance was fine-tuned by testing it on a small section of the image until satisfactory results were obtained. Subsequently, the detection tools were extended to cover the entire drone-captured areas [37]. Within **Figure 9**, a selection of labeled training samples depicting palm trees is displayed, illustrating the outcomes produced by the object detection algorithm employed in the present investigation.

The creation of high-quality training samples plays a crucial role in training deep learning or image classification models. This step is often challenging and time-consuming. In order to equip our deep learning model with the necessary information to accurately identify various crop types in the images, we generated training samples that encompassed different palm trees and other field crops. These samples helped train the model to recognize the size, shape, and spectral signature of these objects. Specifically, the training samples utilized small subimages, referred to as image chips,

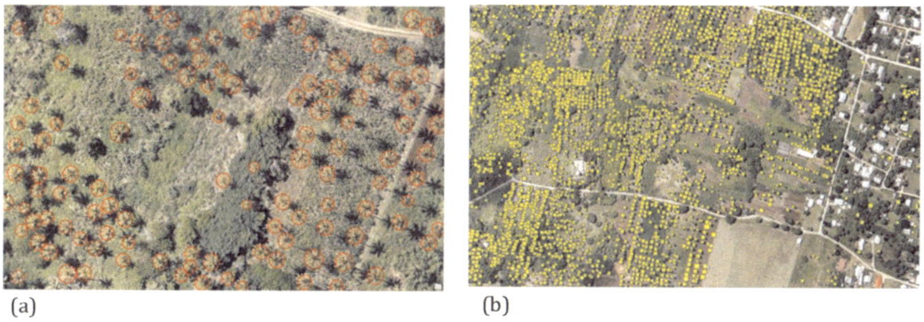

Figure 9.
(a): Recorded training samples for palm trees, (b): Results generated for a larger area utilizing Object Detection Algorithm.

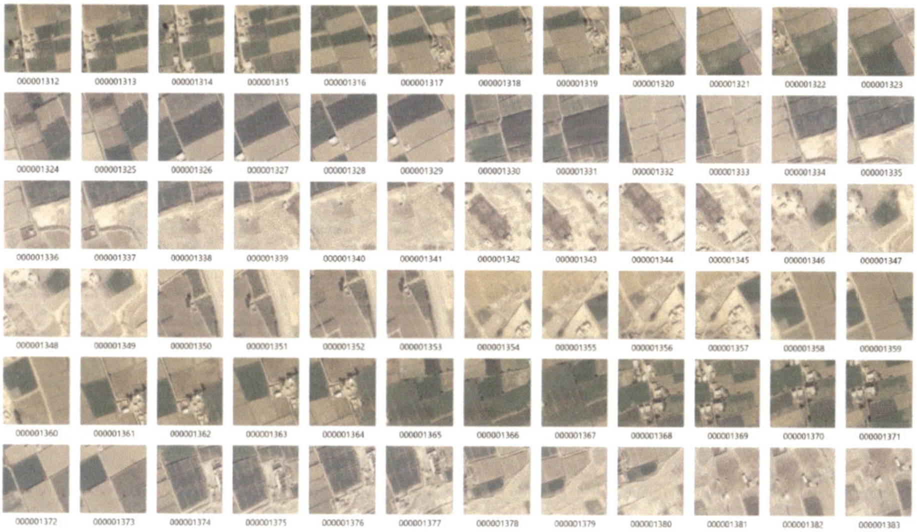

Figure 10.
Recorded training samples for field crops.

that focused on the specific features or classes of focus. **Figure 10** exhibits a subset of the image chips employed within the context of this study [38, 39].

For the chosen workflows (U-net) in this study, the ArcGIS.learn module in the ArcGIS API for Python was utilized. The U-net architecture consists of an encoder network followed by a decoder network. Unlike classification, semantic segmentation requires pixel-level discrimination and feature projection from the encoder [40, 41]. The encoder, shown in **Figure 11**, uses pre-trained classification networks like VGG/ResNet for encoding. The decoder, on the other hand, projects features from the encoder onto higher-resolution pixel space for dense classification using convolution and upsampling.

Tree detection in machine vision relies on human expertise rather than a purely mathematical definition. Deep learning-based object detection differs from other methods by extracting features iteratively, enabling the capture of contextual and global image features for robustness and high accuracy. In this study, a Convolutional Neural Network (CNN) is employed to extract information from high-resolution imagery. The CNN model, depicted in **Figure 12**, includes input, convolution, pooling, fully connected, and output layers. Our ArcGIS Pro model incorporates

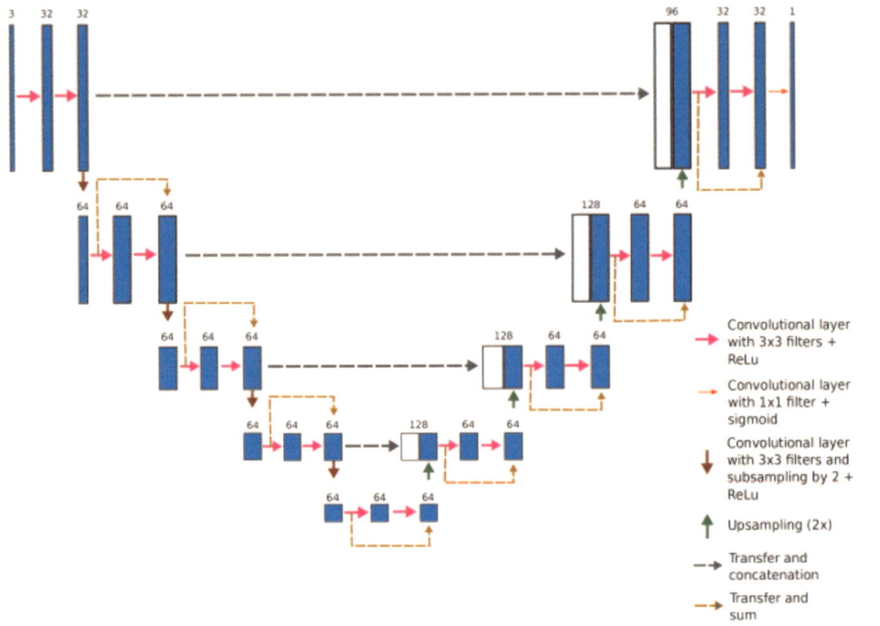

Figure 11.
U-net architecture: Blue boxes represent multi-channel features, white boxes indicate copied feature maps, and colored arrows denote various operations.

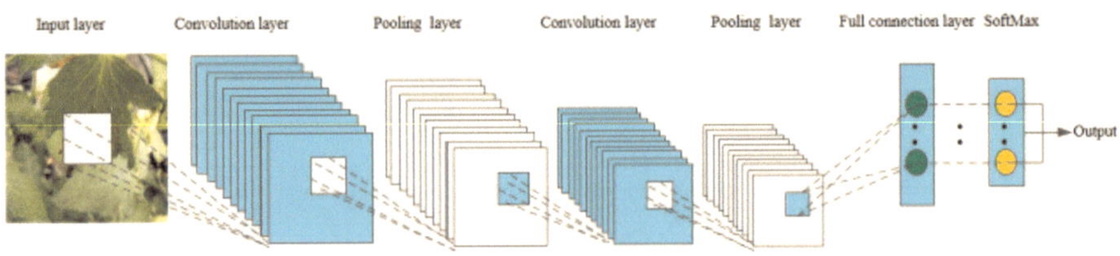

Figure 12.
CNN network layers.

pooling and convolution layers iteratively. CNN's advanced feature extraction handles the challenges of recognizing diverse characteristics in real-world environments, making it a common choice for vegetation cover detection [42].

2.4.3 Data analysis

The main analysis in this study revolves around estimating vegetation health, utilizing the same images employed for deep learning extraction. The assessment of vegetation health involves the calculation of a vegetation health index, specifically the Visible Atmospherically Resistant Index (VARI) [43]. The Vegetation Area Ratio Index (VARI) is employed as an indirect measure of leaf area index (LAI) and vegetation fraction (VF), relying solely on reflectance values within the visible wavelength range.

$$(Rg - Rr)/(Rg + Rr - R(Rg - Rb)) \quad (1)$$

Figure 13.
NDVI results generated from drone imagery (Jabal Ali & Hessyan).

The calculation of the Visible Atmospherically Resistant Index (VARI) involves using reflectance values from the red (Rr), green (Rg), and blue (Rb) bands [43]. Additionally, the estimation of vegetation health utilizes reflectance values from both the visible and near-infrared (NIR) wavelength bands, similar to the normalized difference vegetation index (NDVI). **Figure 13** presents the resulting NDVI for the Hessyan and Jabal Ali communities.

2.4.4 Data evaluation (QA/QC)

To assess the outcomes of vegetation cover extraction and the detection of plant diseases and pests in multispectral drone imagery across Dubai Emirate, several evaluation metrics were employed. These metrics included false negatives (omission errors), false positives (commission errors), detection rate, and an accuracy index (AI) [44]. The AI, which quantifies the balance between omission and commission errors, was computed using the following formula:

$$AI = 100(1 - FP + FN)/REF \tag{2}$$

The terms false positives (FP) and false negatives (FN) represent specific types of errors. FP refers to the instances where a result is falsely identified as positive, while FN pertains to the cases where a positive result is mistakenly classified as negative. In this study, additional evaluation indices were computed, namely Precision, mean Average Precision (mAP), Recall, and the harmonic Mean F1 score. The F1 score is derived from the combination of Precision and Recall defined as follows:

$$Precision = TP/(TP + FP)100\% \tag{3}$$

$$Recall = TP/(TP + FN)100\% \tag{4}$$

In Formulas (3) and (4), TP (True Positives) refers to correctly predicted positive instances, indicating the number of accurately identified lesions by the algorithm. FP (False Positives) represents incorrectly predicted positive instances, indicating the number of lesions inaccurately identified by the algorithm. Conversely, FN (False Negatives) corresponds to incorrectly predicted negative instances that are actually positive. FN represents the count of unacknowledged lesions. The evaluation of detection accuracy is measured by employing the mean Average Precision (mAP) as defined in Formula (6). Initially, the average accuracy for each category within the dataset is computed following the procedure described in Formula (5).

$$Paverage = \sum_{j=0}^{N\ (Class)} Precision\ (j).Recall(j).100\% \tag{5}$$

$$mAP = Paverage/N(class) \tag{6}$$

In the provided formula, N(class) corresponds to the total count of categories. Precision(j) and Recall(j) refer to the precision and recall values specifically associated with class j. The mean Average Precision (mAP) is defined as the average accuracy across all categories. A higher mAP value indicates greater recognition accuracy of the algorithm, while a lower value suggests reduced accuracy. Additionally, the F1 score, which is a significant metric, is employed to assess the accuracy of the deep learning model. The F1 score combines both accuracy and recall, as described in Formula (7).

$$F1 = ((2 * Precision * Recall)/(Precision + Recall)) * 100\% \tag{7}$$

Frames per second (FPS) serves as a metric employed to evaluate the speed at which the deep learning model recognizes information. A greater FPS value indicates a swifter recognition speed of the algorithm, while a lower value suggests a slower recognition speed.

3. Results

In this section, we present the experimental results using real data from 12 communities in Dubai Emirate, encompassing a total area of 770.75 square kilometers. The deep learning models discussed in this paper demonstrated excellent coverage of all crop types, achieving an impressive overall accuracy of 85.4%. Moreover, the detection performance for date palms and GHAF trees was highly promising, with F1-scores of 96.03% and 94.54% respectively.

3.1 Vegetation cover

The results demonstrate that the deep learning models exhibited superior performance compared to the machine learning technique. Specifically, when using RGB color images alone, the deep learning models outperformed the machine learning technique by a minimum of 11%. Moreover, when utilizing NDVI source images, the deep learning models surpassed the machine learning technique by over 28%. Significantly, around 43% of the outcomes obtained from supervised classification were consistent with the deep learning outputs utilizing NDVI. Moreover, there was a notable agreement of 96% between the deep learning results and the photo interpretation method.

During the evaluation of the deep learning model with NDVI, it exhibited marginally better outcomes in comparison to both the deep learning models employing RGB and manual digitization. Subsequent field visits provided further evidence that these

Accuracy criteria	Deep learning using NDVI	Deep learning using RGB	Supervised classification
Sensitivity (Recall)	0.89	0.73	0.54
Specificity	0.95	0.82	0.61
Positive predicted value	0.9	0.83	0.7
Negative predicted value	0.6	0.5	0.3
Prevalence	0.7	0.7	0.5
Detection rate	0.982	0.7	0.6
Detection prevalence	0.9	0.8	0.4
Balanced accuracy	**0.897**	**0.7283**	**0.6154**

Table 6.
Confusion matrix results- overall results for 12 communities.

additional positive results corresponded to plants impacted by prolonged drought, which proved to be difficult to classify solely through photo-interpretation.

Table 6 provides a summary of the confusion matrix results, presenting an overall average for 12 communities and offering an overview of the predicted vegetation cover outcomes using deep learning (with NDVI and Standard RGB), as well as the machine learning (supervised classification) technique. The sensitivity (SN), also known as recall (REC) or true positive rate (TPR), achieved a value of 0.89 for deep learning using NDVI, indicating its suitability for the designated area. The ultimate sensitivity score is 1.0, whereas the worst is 0.0. Conversely, specificity (SP), also referred to as the true negative rate (TNR), quantifies the number of accurate negative predictions, reached an overall value of 0.95 for the deep learning model using NDVI, indicating a relatively high level of accuracy.

The land use classification and object-detection deep learning algorithms achieved high accuracy, with 89.7% for NDVI and 72.8% for RGB. The commission error was minimal, with few instances of including bare soil or grass. Each crop was accurately represented as an individual object. In the Hessyan community, the overall accuracy index reached 87.8%. **Table 7** provides detailed results [45, 46].

Within **Table 8**, a comprehensive compilation of the deep learning outcomes derived from NDVI extraction is presented. The table encompasses essential information such as the count of identified crops, the respective area covered in square kilometers, and the overall Accuracy Index for the chosen communities under

Metric	Deep learning algorithm using NDVI	Reference data (Manual digitization)
True positives	3520 crops	3521 crops
False positives	6 crops	1 crop
False negatives	12 crops	6 crops
Detection rate	98.2%	99.9%
Accuracy index	**87.8%**	**99.9%**

Table 7.
Results of vegetation cover area generated by deep learning using NDVI.

Community	Number of crops	Area SQ KM	Accuracy AI
Saih Shuaib	3117	0.31	97.80%
Hadaeq Sheikh Mohammed Bin Rashid	4841	1.28	89.70%
Aleyas	17,435	1.19	86.60%
Al Kheeran	3007	1.33	83.90%
Al Lesaily	44,433	2.06	97.40%
Margham	32,379	1.33	85.90%
Al Wohoosh	4667	0.03	93.80%
Al Maha	8674	0.84	88.90%
Remah	27,770	0.47	87.50%
Grayteesah	16,049	0.69	84.80%
Al Fagaa	10,298	0.39	92.70%
Hessyan	3520	0.60	87.80%
Total/Overall	**176,190**	**10.53**	**89.73%**

Table 8.
Outcomes of deep learning utilizing NDVI extracted from drone imagery.

investigation. Notably, the communities of Al Lesaily, Margham, and Remah exhibited the highest prevalence of recorded crops, indicating their significant vegetation presence across Dubai Emirate.

Figure 14 showcases sample outcomes obtained through the deep learning-based land use classification model using NDVI images. Thorough preparation of every source image was diligently carried out, as extensively discussed in this paper. The creation of a vector layer representing vegetation cover was achieved through a

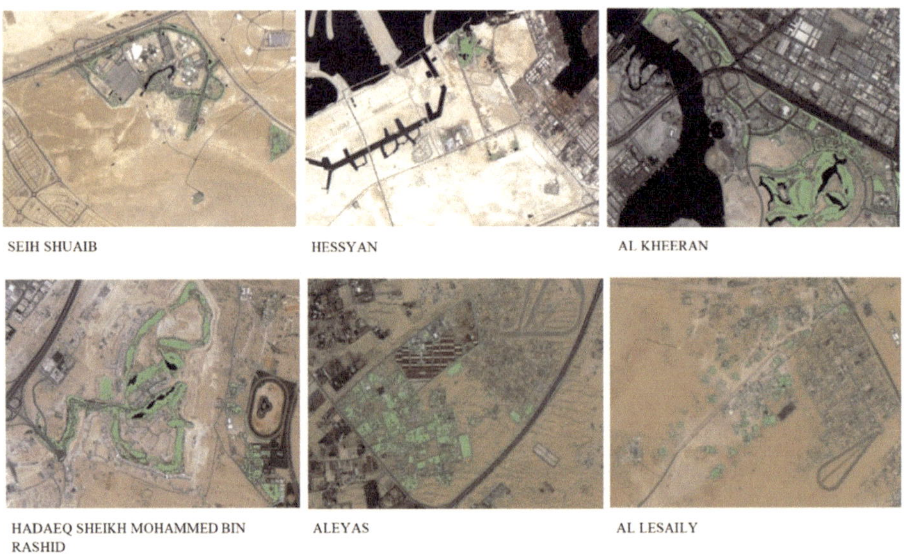

Figure 14.
Results of NDVI based deep learning model for vegetation extraction from UAS multispectral imagery.

comprehensive methodology that incorporated advanced geospatial analysis and deep learning techniques. The results showcased in this study portray trees, shrubs, or grass through green pixels, barren land through silver color, and urban areas depicted in various shades of blue. Notably, Seih Shuaib exhibits the highest accuracy index at 97.8%, attributed to the utilization of hydroponic technology for cultivating in-demand micro-greens and herbs in several farms. Conversely, Al Kheeran demonstrates the lowest accuracy index at 83.9%. The overall average accuracy index for all communities within the study's scope is calculated to be 89.73%.

3.2 Date palms and GHAF trees detection

The tree detection model's accuracy was evaluated by comparing it with photo-interpreted drone images. The deep learning algorithm outperformed the machine learning method in detecting date palms and GHAF trees. The deep learning model accurately predicted 336 trees, while the supervised classification predicted 289 trees. Overall, the deep learning model showed superior performance, as evidenced by F1 scores. **Figure 15** provides an example of detected trees in different locations [47].

Both methods demonstrated satisfactory classification results for GHAF trees and date palms, achieving accuracies of over 79%. However, the deep learning object-based method outperformed in accurately detecting the target trees, with overall accuracies surpassing 95% for GHAF trees and 97% for date palms. **Figure 16** provides a visual representation showcasing the disparities observed in the detection procedures, emphasizing a higher precision of approximately 16% for GHAF trees and 13% for date palms.

The initial assessment revealed that errors in tree detection occurred primarily when palm trees were obscured by other tree canopies or when trees with similar physical characteristics to palm trees were present, particularly coconut trees. These errors were relatively minor and mainly observed in areas where coconut trees were planted. Additionally, the detectability of palm trees was affected in cases where the trees were located at the image edges, resulting in some parts of the crown area extending across two images. Detection errors also arose from the size of the crown, particularly in young palms with smaller crown sizes. This discrepancy can be attributed to the limited inclusion of young palm samples with small crowns, as the study

(a) (b)

Figure 15.
(a): Detected date palm trees in Al KIFAF road; (b): Detected GHAF trees in Hadaeq Sheikh Mohammed Bin Rashid.

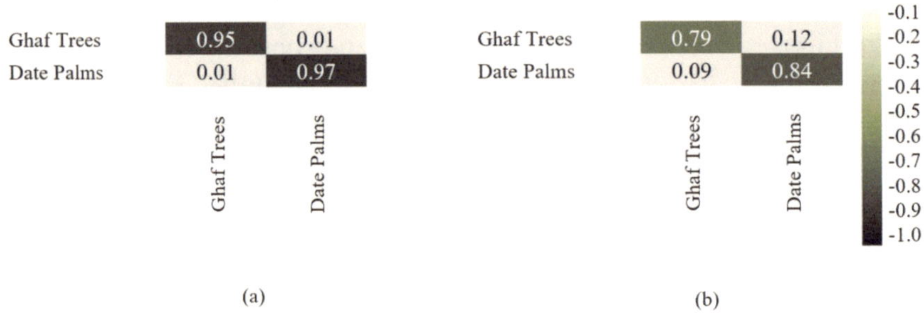

Figure 16.
Confusion matrix for date palm trees and GHAF trees under different scenarios:(a) deep learning, (b) supervised classification.

primarily focused on mature palms prevalent within the study area. However, addressing the challenges associated with crown size errors could be accomplished by incorporating a greater number of young palm samples with small crowns into the training data, thereby improving the model's performance.

3.3 Performance comparisons

Tables 6 and **9** summarize the performance of deep learning models using NDVI and RGB imagery, compared to supervised classification. The NDVI-based deep learning model outperforms the RGB-based model. Both deep learning models outperform supervised classification in vegetation cover classification and tree object detection. The NDVI-based model shows slightly better results. Deep learning is more accurate but requires longer training time. In areas with low vegetation cover, supervised classification is acceptable. For simplicity and reduced hardware dependency, use supervised classification when complex features are absent.

4. Conclusions

The combination of Unmanned Aircraft Systems (UAS) and deep learning object detection methods enables accurate crop identification and productivity analysis, achieving an overall accuracy of 89.7%. Specifically, the detection rates for date palm trees and GHAF trees reach 96.03% and 94.54%, respectively. The implementation of this approach in Dubai Municipality has demonstrated its potential in addressing

Metric	Date palms (Deep learning)	Date palms (Supervised classification)	GHAF trees (Deep learning)	GHAF trees (Supervised classification)
True positives	10,663	9238	9592	7983
Precision	97.3%	84.3%	95.4%	79.4%
Recall	94.8%	86.8%	93.7%	77.2%
F1 score	96.03%	85.5%	94.54%	78.28%

Table 9.
Comparison of detected trees by method and type.

agricultural challenges by providing up-to-date, high-quality data for informed decision-making.

Precision farming, which integrates sensor data, imaging, and real-time analytics, plays a crucial role in enhancing farm productivity through the mapping of spatial variability in fields. The data collected through drones during this work serves as a valuable resource for activating analytical models in agriculture. By supporting precision farming practices, UAS facilitate soil health analysis, irrigation planning, yield estimation, fertilizer application, and weather analysis. The combination of spatial data from drones with other data sources and analytic solutions generates actionable information for agricultural management.

The Normalized Difference Vegetation Index (NDVI) measured at different crop stages exhibits a robust correlation with crop yield. Monitoring crop growth at critical stages using NDVI maps, alongside other indexes like the Crop-Water Stress Index (CWSI) and Canopy Chlorophyll Content Index (CCCI), enables accurate estimation of crop yield and early identification of issues. Multispectral drone imagery, equipped with sensors such as infrared and hyperspectral, proves to be an effective method for detecting plants under stress, differentiating between crops, and assessing crop health in agricultural mapping tools.

Acknowledgements

This work was supported by the Geographic Information Systems Centre (GISC), Dubai Municipality.

Author details

Lala El Hoummaidi[1]*, Abdelkader Larabi[2] and Khan Alam[3]

1 Statistics Centre of Abu Dhabi, Abu Dhabi, United Arab Emirates

2 Laboratory of Analysis and Modeling of Water and Natural Resources (LAMERN), Mohammed V University in Rabat, Morocco

3 Department of Physics, University of Peshawar, Khyber Pakhtunkhwa, Pakistan

*Address all correspondence to: lala.elhoummaidi@gmail.com

References

[1] Kekane MA. Indian agriculture-status, importance and role in Indian economy. International Journal of Agriculture and Food Science Technology. 2013;**4**(4):343-346

[2] Fan M, Shen J, Yuan L, Jiang R, Chen X, Davies WJ, et al. Improving crop productivity and resource use efficiency to ensure food security and environmental quality in China. Journal of Experimental Botany. 2011;**63**(1): 13-24. DOI: 10.1093/jxb/err248

[3] Oyakhilomen RGZ. Agricultural production and economic growth in Nigeria: Implication for rural poverty alleviation. Quarterly Journal of International Agriculture. 2014;**53**(3): 207-223. DOI: 10.22004/ag.econ.195735

[4] Awokuse TO. Does agriculture really matter for economic growth in developing countries? In: The American Agricultural Economics Association Annual Meeting. Agricultural and Applied Economics Association. Milwaukee, Newark, USA; 2009. DOI: 10.22004/ag.econ.49762

[5] Badiene O. Sustaining and accelerating Africa's agricultural growth recovery in the context of changing global food prices. IFPRI Policy Brief. 2008;**9**:1-4

[6] de Gennaro BC, Forleo MB. Sustainability perspectives in agricultural economics research and policy agenda. Agricultural Economics. 2019;**17**:7. DOI: 10.1186/s40100-019-0134-8

[7] Food and Agriculture Organization of the United Nations. The Future of Food and Agriculture: Trends and Challenges. Rome: FAO; 2017

[8] Balogh JM, Jámbor A. The environmental impacts of agricultural trade: A systematic literature review. Sustainability. 2020;**12**:3. DOI: 10.3390/su12031152

[9] Food and Agriculture Organization of the United Nations. The State of Agricultural Commodity Markets. Agricultural Trade, Climate Change and Food Security. Rome, Italy: FAO; 2018

[10] Garsous G. Trends in policy indicators on trade and environment. In: OECD trade and environment working papers. Paris, France: OECD; 2019. DOI: 10.1787/18166881

[11] Kwan C, Gribben D, Ayhan B, Bernabe S, Plaza A, Selva M. Improving land cover classification using extended multi-attribute profiles (EMAP) enhanced color, near infrared, and LiDAR data. Remote Sensing. 2020;**12**: 1392. DOI: 10.3390/rs12091392

[12] Tan K, Zhang Y, Wang X, Chen Y. Object-based change detection using multiple classifiers and multi-scale uncertainty analysis. Remote Sensing. 2019;**11**:359. DOI: 10.3390/rs11030359

[13] Van der Meij B, Kooistra L, Suomalainen J, Barel J, De Deyn G. Remote sensing of plant trait responses to field-based plant-soil feedback using UAV-based optical sensors. Biogeosciences. 2017;**14**:733-749. DOI: 10.5194/bg-14-733-2017

[14] Zare A, Bolton J, Gader P, Schatten M. Vegetation mapping for landmine detection using long-wave hyperspectral imagery. IEEE Transactions on Geoscience and Remote Sensing. 2007;**46**:172-178. DOI: 10.1109/TGRS.2007.906438

[15] Wellmann T, Lausch A, Andersson E, Knapp S, Cortinovis C,

Jache J, et al. Remote sensing in urban planning: Contributions towards ecologically sound policies? Landscape and Urban Planning. 2020;**204**:2-10. DOI: 10.1016/j.landurbplan.2020.10392

[16] Skarlatos D, Vlachos M. Vegetation removal from UAV derived DSMS, using combination of RGB and NIR imagery. ISPRS Annals of the Photogrammetry, Remote Sensing and Spatial Information Sciences, Volume. 2018;**IV-2**:255-262. DOI: 10.5194/isprs-annals-IV-2-255-2018

[17] Hellesen T, Matikainen L. An object-based approach for mapping shrub and tree cover on grassland habitats by use of LiDAR and CIR orthoimages. Remote Sensing. 2013;**5**:558-583. DOI: 10.3390/rs5020558

[18] Ayhan B, Kwan C, Kwan L, Skarlatos D, Vlachos M. Deep learning models for accurate vegetation classification using RGB image only. In: Proceedings of the Geospatial Informatics X (Conference SI113). Proceedings of the SPIE. Anaheim, CA, USA; 2020. DOI: 10.1117/12.2557833

[19] Guirado E, Tabik S, Alcaraz-Segura D, Cabello J, Herrera F. Deep-learning versus OBIA for scattered shrub detection with Google earth imagery: Ziziphus lotus as case study. Remote Sensing. 2017;**9**:1220. DOI: 10.3390/rs9121220

[20] Yang L, Wu X, Praun E, Ma X. Tree detection from aerial imagery. In: Proceedings of the 17th ACM SIGSPATIAL International Conference on Advances in Geographic Information Systems. Seattle, WA, USA; 2009. pp. 131-137

[21] Snehal SS, Sandeep SV. Agricultural crop yield prediction using artificial neural network approach. International Journal of Innovative Applications of Artificial Intelligence in Agriculture Research in Electrical, Electronics, Instrumentation and Control Engineering. 2014;**2**(1):683-686. DOI: 10.1.1.429.1195

[22] Zhang X, Han L, Han L, Zhu L. How well do deep learning-based methods for land cover classification and object detection perform on high resolution remote sensing imagery? Remote Sensing. 2020;**12**:417. DOI: 10.3390/rs12030417

[23] Song H, He Y. Crop nutrition diagnosis expert system based on artificial neural networks. In: 3rd International Conference on Information Technology and Applications. Sydney, Australia; 2005. DOI: 10.1109/ICITA.2005.108

[24] Papageorgiou EI, Markinos AT, Gemtos TA. Fuzzy cognitive map-based approach for predicting crop production as a basis for decision support system in precision agriculture application. Applied Soft Computing. 2011;**11**(4):3643-3657. DOI: 10.1016/j.asoc.2011.01.036

[25] Dai X, Huo Z, Wang H. Simulation of response of crop yield to soil moisture and salinity with artificial neural network. Field Crops Research. 2011;**121**(3):441-449. DOI: 10.1016/j.fcr.2011.01.016

[26] Rehman A, Jingdong L, Khatoon R, Hussain I. Modern agricultural technology adoption its importance, role and usage for the improvement of agriculture. American-Eurasian Journal of Agricultural & Environmental Sciences. 2016;**16**(2):284-288. DOI: 10.5829/idosi.aejaes.2016.16.2.12840

[27] Purkis S, Riegl B. Geomorphology and Reef Building in the SE Gulf. 2012. DOI: 10.1007/978-94-007-3008-3_3

[28] Bolleter J. Desert paradises: Surveying the landscapes of Dubai's urban model. Taylor & Francis. 2019. DOI: 10.4324/9781351129763

[29] Fathelrahman E, Gheblawi M, Muhammad S, Dunn E, Ascough J, Green T. Optimum returns from greenhouse vegetables under water quality and risk constraints in the United Arab Emirates. Sustainability. 2017;**9**(5): 719. DOI: 10.3390/su9050719

[30] Shahmoradi J, Talebi E, Roghanchi P, Hassanalian M. A comprehensive review of applications of drone technology in the mining industry. Drones. 2020;**4**:3. DOI: 10.3390/drones4030034

[31] Christiansen MP, Laursen MS, Jørgensen RN, Skovsen S, Gislum R. Designing and testing a UAV mapping system for agricultural field surveying. Sensors. 2017;**17**:2703. DOI: 10.3390/s17122703

[32] Starý K, Jelínek Z, Kumhálová J, Chyba J, Balážová K. Comparing RGB - Based vegetation indices from UAV imageries to estimate hops canopy area. Agronomy Research. 2020;**18**:4. DOI: 10.15159/ar.20.169

[33] Pauly K. Towards Calibrated Vegetation Indices from UAS-derived Orthomosaics. 2016. DOI: 10.13140/RG.2.2.21842.35524

[34] Klaas P. Applying conventional vegetation vigor indices to UAS-derived Orthomosaics: Issues and considerations. In: 12th International Conference for Precision Agriculture. Sacramento (CA, USA); 2014

[35] Turner D, Lucieer A, Watson C, Turner D, Lucieer A, Watson C. An automated technique for generating Georectified mosaics from ultra-high resolution unmanned aerial vehicle (UAV) imagery, based on structure from motion (SfM) point clouds. Remote Sensing. 2012;**4**:1392-1410. DOI: 10.3390/rs4051392

[36] Höhle J. Generating topographic map data from classification results. Remote Sensing. 2017;**9**:3. DOI: 10.3390/rs9030224

[37] Du Z, Yang J, Ou C, Zhang T. Smallholder crop area mapped with a semantic segmentation deep learning method. Remote Sensing. 2019;**11**:888. DOI: 10.3390/rs11070888

[38] Najafabadi MM, Villanustre F, Khoshgoftaar TM, et al. Deep learning applications and challenges in big data analytics. Journal of Big Data. 2015;**2**:1

[39] Frank E-S, Zhen Y, Han F, Shailesh T, Matthias D. An introductory review of deep learning for prediction models with big data. Frontiers in Artificial Intelligence. 2020;**3**(3):4. DOI: 10.3389/frai.2020.00004

[40] Lamba H. Understanding semantic segmentation with UNET, A salt identification case study. Towards Data Science. 2019

[41] Michelle L, Jana R, Utku AO, Aziz TA, Marie AE, Tabea K, et al. A U-net deep learning framework for high performance vessel segmentation in patients with cerebrovascular disease. Frontiers in Neuroscience. 2019;**13**:97. DOI: 10.3389/fnins.2019.00097

[42] Türkoğlu M, Hanbay D. Plant disease and pest detection using deep learning-based features. Turkish Journal of Electrical Engineering and Computer Sciences. 2019;**27**:1636-1651. DOI: 10.3906/elk- 1809-181

[43] Liu YH. Feature extraction and image recognition with convolutional

neural networks. Journal of Physics: Conference Series. 2018;**1087**:6. DOI: 10.1088/1742-6596/1087/6/062032

[44] Mishra A. Metrics to evaluate your machine learning algorithm. Towards Science. 2018

[45] Sogawa T, Tabuchi H, Nagasato D, Masumoto H, Ikuno Y, Ohsugi H, et al. Accuracy of a deep convolutional neural network in the detection of myopic macular diseases using swept-source optical coherence tomography. PLoS One. Apr 16 2020;**15**(4):e0227240. DOI: 10.1371/journal.pone.0227240. PMID: 32298265; PMCID: PMC7161961

[46] Zhao H, Yang C, Guo W, Zhang L, Zhang D. Automatic estimation of crop disease severity levels based on vegetation index normalization. Remote Sensing. 2020;**12**:12. DOI: 10.3390/rs12121930

[47] Yarak K, Witayangkurn A, Kritiyutanont K, Arunplod C, Shibasaki R. Oil palm tree detection and health classification on high-resolution imagery using deep learning. Agriculture. 2021; **11**:2. DOI: 10.3390/agriculture11020183